UNIVERSITY OF STRATHCLYDE
30125 00488450 7

KV-056-640

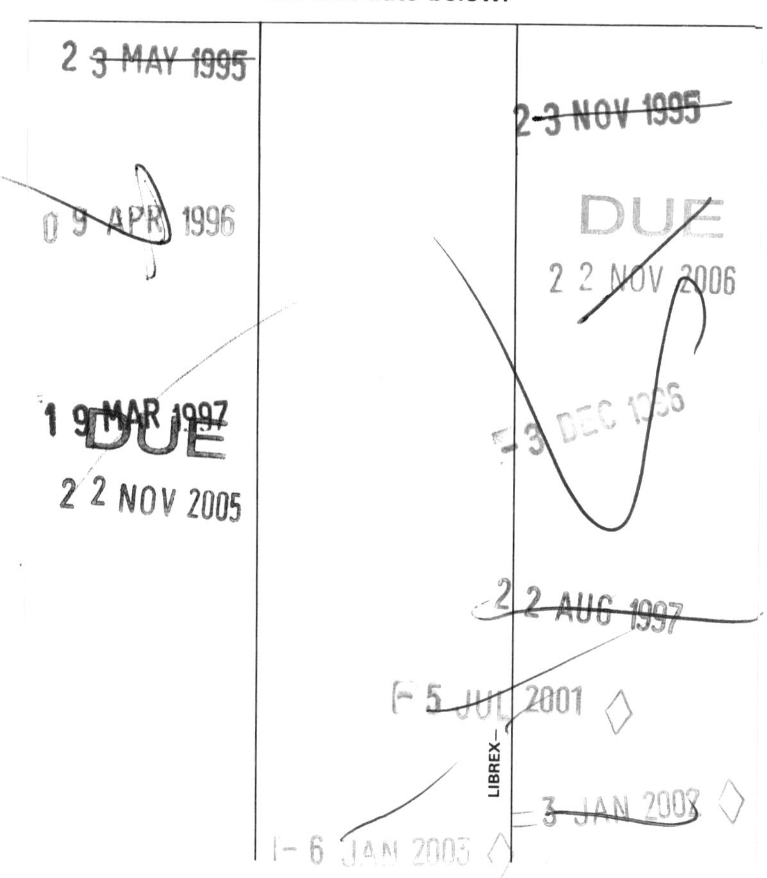
23 MAY 1995
09 APR 1996
19 MAR 1997
DUE
22 NOV 2005
23 NOV 1995
DUE
22 NOV 2006
3 DEC 1996
22 AUG 1997
5 JUL 2001
LIBREX
3 JAN 2003
6 JAN 2003

Practical Programmable Circuits

Practical Programmable Circuits

A Guide to PLDs, State Machines, and Microcontrollers

James D. Broesch
Science Applications International Corporation
San Diego, California

Academic Press, Inc.
Harcourt Brace Jovanovich, Publishers
San Diego New York Boston
London Sydney Tokyo Toronto

This book is printed on acid-free paper. ♾

Academic Press, Inc.
San Diego, California 92101

United Kingdom Edition published by
Academic Press Limited
24–28 Oval Road, London NW1 7DX

Library of Congress Cataloging-in-Publication Data

Broesch, James D.
Practical programmable circuits : a guide to PLDs, state machines, and microcontrollers / James D. Broesch.
p. cm.
Includes index.
ISBN 0-12-134885-7
1. Programmable logic devices. 2. Logic design. I. Title.
TK7872.L64B76 1991
621.39'5--dc20 916588
CIP

PRINTED IN THE UNITED STATES OF AMERICA
91 92 93 94 9 8 7 6 5 4 3 2 1

This book is dedicated to my wife Lynn
and to my three very special children
Valerie, David, and Michael

Contents

Preface xi

Acknowledgments xv

1

An Introduction to Programmable Logic 1

1.1 The Programmable Logic Design Environment 2
1.2 The Programmable Logic Development Cycle 3
1.3 Some Special Concerns with Programmable Logic 7
1.4 Getting Started 9
1.5 Chapter Summary 10

2

Basic Logic Design 12

2.1 Logic Design 13
2.2 Selected Topics in Basic Logic 16
2.2.1 Combinatorial Logic 18
2.2.2 State Logic 29
2.3 Chapter Summary 34

3

Combinatorial PLDs 36

3.1 Programmable Logic and the PROM 39
3.2 Programmable Array Logic 47
3.3 Programmable Logic Arrays 56
3.4 A Typical PLD Application 59
3.5 PROM, PAL, and PLA Comparisons 63
3.6 Chapter Summary 65

4

State Machines 67

4.1 An Introduction to State Machines 67
4.2 Classic State Machines 69
4.3 Chapter Summary 86

5

Software Development 89

5.1 Introduction to PLD Software 90
5.2 Specific Software Packages 94
5.2.1 PALASM 94
5.2.2 AMAZE 94
5.2.3 CUPL 95
5.2.4 ABEL 95
5.2.5 PLDesigner 95
5.2.6 PLAN II 96
5.2.7 Device Specific Packages 96
5.3 The proLogic Compiler 97
5.3.1 Pin Names and Signal Conventions 99
5.3.2 Truth Tables 100
5.3.3 State Diagrams 100
5.3.4 PAL Version of the Four Bit Counter 104
5.3.5 Simulation 108
5.4 Miscellaneous Comments on Software 111
5.5 Chapter Summary 112

6

Advanced Forms of PLDs 114

6.1 The PAL22V10 115
6.2 PSG-506/507 123
6.3 ASICs and Third Generation PLDs 130
6.4 Altera 131
6.5 Xilinx 138
6.6 PLDs as Competition for Gate Arrays 142
6.7 GALs 143
6.8 Chapter Summary 144

7

General PLD Design Issues 146

7.1 Philosophy of Programmable Device Design 146
7.2 Design for Testability 148

7.3 Metastability 154
7.4 High-Speed Circuit Design 156
7.5 Security 162
7.6 Chapter Summary 163

8

Variations on the Theme 164

8.1 Microsequencers 164
8.2 RISC versus CISC 174
8.3 Writable Control Stores 175
8.4 Chapter Summary 176

9

Introduction to Microcontrollers 177

10

Hardware Architecture of Microcontrollers 181

10.1 Basic Features 181
10.2 Common Optional Features 185
10.3 Exotic Optional Features 188
10.4 Four Bit Units 189
10.5 Eight Bit Units 190
10.6 Sixteen Bit Units 192
10.7 I/O Interfacing 192
10.8 Chapter Summary 199

11

Microcontrollers and Software 200

11.1 Patterns, Microcode, and Object Code 200
11.2 Elementary Instructions 203
11.3 Simple Program Structure 207
11.4 Subroutines and Program Structure 215
11.5 Interrupts 218
11.6 Real Time Multi–Tasking 220
11.7 Chapter Summary 233

12

Additional Tools of the Trade 234

12.1 Basic Tools 234
12.2 Logic Analyzers 235
12.3 Monitors 237
12.4 Simulators 241
12.4.1 Built-in Simulators 241
12.4.2 General Purpose Simulators 242
12.4.3 Device Specific Simulators 243
12.5 In-Circuit Emulators 244
12.6 Chapter Summary 246

13

A Guide to Choosing Programmable Circuits 248

13.1 Proof-of-Concept Phase 249
13.2 Preproduction Units 250
13.3 Production Systems 252
13.4 Chapter Summary 253

14

Conclusion 254

Appendix A

An Arcane History of a Few Acronyms 259

Appendix B

Data Sheets 262

Appendix C

References and Sources 276

Glossary 278

Index 285

Preface

Programmable logic circuit design can be defined as the art of applying software techniques to hardware design problems. The purpose of this book is to provide a guide to both the hardware and software techniques required to successfully design circuits using programmable logic. This is no small undertaking. The successful, cost-effective solution to a design problem may require the use of a simple combinatorial chip, a state machine, or a microcontroller. Much has been written about these areas individually. A comprehensive treatment, however, comparing and contrasting the techniques from a practical point of view has not been available until now.

This book is oriented toward, but not limited to, the working digital design engineer. It will be of use to many other people as well. The number of people interested in programmable logic design is increasing every day. The list includes analog engineers looking to diversify their skills; management personnel who need to understand and control design projects; and newly graduated engineers who have not been exposed to the practical aspects of using programmable logic. Instrumentation scientists, technicians, and students are also developing an increased interest in programmable logic.

Many factors have contributed to the increasing interest in programmable logic design techniques. Short design cycles, fast product introduction, strong competition, and compressed product life cycles all make the selection and appropriate use of programmable logic mandatory for many successful designs.

Specifically, this text is designed to provide an expert guide to the subject of programmable logic design. This is not to say that one must be an expert to make use of this book. In fact, quite the opposite. This book is designed to efficiently impart the general information, rules, and guidelines used by an expert in programmable logic. As such, it will be of use both to the experienced designer and to those who wish to expand their understanding of programmable logic systems.

Expertise in any given field is demonstrated by the understanding of a wide range of ideas and techniques in the given discipline. Thus we cover neglected topics in both very simple areas and in highly sophisticated ones as well. For example, this book spends more time dealing with the basic structure of PLDs than is found in the manufacturer's documentation. The information on basic structure may be review material for the experienced digital designer, but it will be of special value to those who are new to the concepts of programmable logic design. At the other extreme, we discuss the concepts of programming efficiently in a multitasking environment. This is usually a fairly advanced topic, even for computer science majors. Nevertheless, understanding the concepts of multi-tasking is important for the digital designer, since many of the new microcontrollers are designed to support this capability. Few digital design engineers are sufficiently cognizant of the advanced software techniques to make use of these capabilities.

An expert in a field must be able to explain the reasoning behind a particular conclusion. It is not enough to simply conclude, for example, that a microcontroller should be used instead of a PAL for a given situation. One must be able to explain why the approach chosen is the most cost-effective solution to the problem. To aid in both the evaluation of the various techniques and in explanation of the decision to a critical audience, most chapters conclude with a simple summary of the main concepts covered.

The text is organized into three main areas. The first area of concentration is an overview of programmable logic. This covers the basic theory behind, and the architectures of, the more common devices. The second area of concentration is advanced PLDs and microsequencers. These circuits are required when complex designs requiring sequences of operations are needed. The final area covered is the microcontroller. These flexible devices bring truly amazing processing power down to the chip level.

Each chapter focuses on a specific aspect of these three major areas. Chapter 1 provides a general discussion of the design environment for working with programmable logic. The types of computers and other hardware support used in the design of programmable logic are discussed. The general procedures used in the development process are introduced.

Chapter 2 is a quick review of fundamental concepts of logic. The material will be a useful refresher for those readers with several intervening years between their education in logic and the reading of this book. The main goal in Chapter 2 is to present the mathematical background necessary to understand the architectures of the various devices discussed later in the book.

Chapter 3 introduces the three basic PLD architectures: the PROM, the PAL, and the PLA. The theory behind the basic architectures is presented, along with the relative merits of each architecture.

Chapter 4 discusses the theory and design of state machines. The fundamental ideas behind the classic state machines, as well as standard nomenclature, are covered.

With a good grounding in basic logic theory and state machine design, we turn our attention to the specifics of programming PLDs in Chapter 5. Specific software development tools are discussed. PLD assemblers and microcode assemblers (also known as meta-assemblers) are examined.

In Chapter 6 we examine the more advanced "programmable gate arrays." These sophisticated devices are becoming more popular by the day. The major architectures provided by various vendors are presented.

Chapter 7 is a smorgasbord of often-neglected topics on PLDs. The general philosophy of designing with PLDs, design for testability, high speed layout techniques for fast switching devices, and other such topics are covered.

Chapter 8 looks at microsequencers. The basic architecture and the role of microsequencers in the ongoing RISC versus CISC debate are introduced.

Chapters 9, 10, and 11 cover microcontrollers. Several devices, from simple 4 bit units to sophisticated 16 bit units, are examined. Particular attention is paid to software design techniques. Microcontrollers used to perform complex logic functions, particularly in automotive-type applications, require sophisticated interrupt structures and multi-tasking programs. Few hardware design engineers get the necessary software exposure in school to fully understand and make use of these software techniques. With this in mind, the section on software development for microcontrollers covers interrupts, multi-tasking, and other sophisticated subjects.

The concluding chapters of the book provide additional guidelines on selection criteria for the various programmable logic techniques. We also take a look at new trends and future directions that programmable logic is likely to take.

Finally, this book is a practical text. We do not spend great effort in proving logic theorems or trying to present the latest theoretical algorithms for network switching or logic reduction. Not that these things are not important, but this book is intended to provide a broad spectrum of information for the working engineer using programmable logic. Many professional and academic journals cover the latest theoretical developments, but a comprehensive discussion of the field of programmable logic has not been available previously. Hopefully, this text will fill the need for a simple, comprehensive discussion of the techniques for practical design using programmable logic.

Acknowledgments

The completion of any work that covers the scope that this book does is possible only through the efforts of many people. While it is not possible to list all those that contributed to this project in some way, I would like to thank some of them in particular.

I am indebted to Dr. J.R. Beyester, Founder and CEO of Science Applications International Corporation (SAIC), for his support of this project and also for providing an employee-owned company rich in opportunities. I would also like to acknowledge my appreciation of the efforts of SAIC's Tom Hicks, Bill Scott, and Larry Hamerman.

Many technical and support people from various organizations helped in providing information and reprint permission. Dave Allen (MINC), Susan Cain (Altera), Dave Crumpton (Motorola), Jim Deines (InLab), Dale Prull (WSI), Andres Martinez (Motorola), and Brandon Nixon (TI) all helped with the gathering of information presented here.

Finally, Riley Woodson's work in reviewing, critiquing, and checking the manuscript has been invaluable. Riley's many suggestions for improvements have made this a better book than it would otherwise have been.

I am also indebted to Craig Kalas for his review and critique of the final draft of the manuscript.

1

An Introduction to Programmable Logic

Programmable logic design is claiming an ever larger portion of modern system design activity. Programmable logic offers solutions that are either nonexistent or prohibitively expensive if implemented with other design approaches.

The vehicles that are used to implement the design solutions can take a variety of shapes. On one end of the spectrum are the programmable logic devices (PLDs). These are integrated circuits (ICs) with a collection of logic gates that are electronically wired together under software control. Somewhere in the middle of the spectrum are state machines. These circuits make use of programmable read only memories (PROMs) (or PLDs) and latches to provide a very orthogonal environment for executing sequential logic functions. At the high end of the spectrum are microcontrollers. Microcontrollers are small computers fabricated on a single IC. These computers can be programmed, often in ways similar to their full size cousins, to accomplish a wide variety of logic and control functions.

Linking all of these devices is the idea of programmability. Why is this such a fundamental issue? The answer lies in the efficiency and flexibility that programmability provides. Programmable devices allow the hardware design to proceed even when the details of the functionality are still in a formative phase. This is often critical to meeting aggressive product introduction cycles. Changes that would require extensive redesign in a hard-wired system can be accommodated with simple changes to the logic equations in a system based on programmable logic. Changes to the programming are often simpler and faster than rewiring a design.

An additional advantage of programmability is the cost-effectiveness of these flexible parts. It is not necessary to stock a wide variety of parts when programmable logic is baselined in the design. A smaller number of parts can fill a larger number of functions. This makes stocking easier, makes manufacturing less vulnerable to supply interruptions, and allows cost discounting on volume purchasing of programmable parts.

Programmable logic may be the only practical way to effectively implement critical designs. Very tight printed circuit boards require that every chip do as much as possible. Designing custom or semicustom ICs is one approach to packing as much functionality as possible into a chip. This is an expensive and time-consuming process, however. Programmable logic provides a simple, cost-effective alternative for efficiently making use of available printed circuit board real estate.

The use of programmable logic has brought many changes to the electronic engineering field. The digital designer is not the only one affected. The analog design engineer is finding that "pure" analog designs are becoming an endangered species. Today, compact discs are the preferred media for music, and phone transmissions are not only digitally switched but are actually transmitted digitally. Radios are no longer tuned by analog oscillators but rather by digitally synthesized frequencies. The introduction of high definition television (HDTV), in whatever form predominates, will blur the line between analog and digital even further. These changes make it harder for the analog engineer to function without a good understanding of digital techniques. This is particularly true of the role of programmable logic in traditionally analog products, such as consumer radios.

The changes have not left the project managers and technical leads unaffected either. Lead staff who have not been actively involved in the day-to-day work with the new devices and products coming on the market will find this book a valuable and concise reference. Controlling product development costs requires both accurately estimating the level of effort for tasking and controlling unanticipated cost drivers. Cost control of modern designs can be nearly impossible if one does not understand both the advantages and the risk areas involved in designing with programmable logic. The options for contingency planning are one of the particularly strong points of programmable logic.

1.1 The Programmable Logic Design Environment

The basic idea behind all programmable logic is to take an off-the-shelf integrated circuit and "add in" custom capabilities to solve unique design problems. The value of this "value added" step depends upon the understanding and skill of the designer.

The understanding and competent use of the various device families and design techniques is only the first step in effectively meeting the challenges of using programmable logic. It is not enough for the engineer to be able to choose between using a complex PLD and a simple microcontroller on purely technical grounds. He or she must also consider other tangibles such as the cost of programming equipment and the availability and sophistication of software and debugging tools as well as intangibles such as learning curves and user proficiency.

All programmable logic has several things in common. Programmable logic is based on the use of computers to develop the codes used to program the logic devices. The computer is usually called a development station and can be as simple as a PC or as sophisticated as a minicomputer. The process of developing programmable logic on the development station is similar regardless of whether a PLD, state machine, or microcontroller is being developed.

As computers have become a major design tool, the profession of design engineer has changed. Design teams have been reduced in size considerably. The designer rarely sits down with pencil and paper with the sole intention of turning a sketch over to a team of drafters, technicians, and printed circuit designers. The modern design environment requires the designer to be an interactive part of a smaller, more cohesive team. The computer is the central point of this new environment. Designs are specified, entered, simulated, evaluated, and modified in less time than was previously allocated for the first pass design task alone.

Modern hardware designers are often both consumers and producers of software. The designer is obviously a consumer when making use of the various software tools on the development stations. The designer is a producer of software when developing the source files which are used in the implementation of programmable logic devices. Unfortunately, many engineers are not well prepared for the development, documentation, and maintenance of the software portion of the design. One of several key goals of this book is to help provide a good grounding in the software aspects of this essentially hardware-oriented environment.

1.2 The Programmable Logic Development Cycle

Figure 1-1 is a flow chart of the programmable logic development cycle. The first step in the cycle is the specification of the requirements and the functionality of the logic. This specification can be as simple as "we need a decoding block here." Or it can be a detailed mathematical representation of precisely what logic functions are anticipated. A major advantage of programmable logic is the flexibility it allows during the early phases of the design.

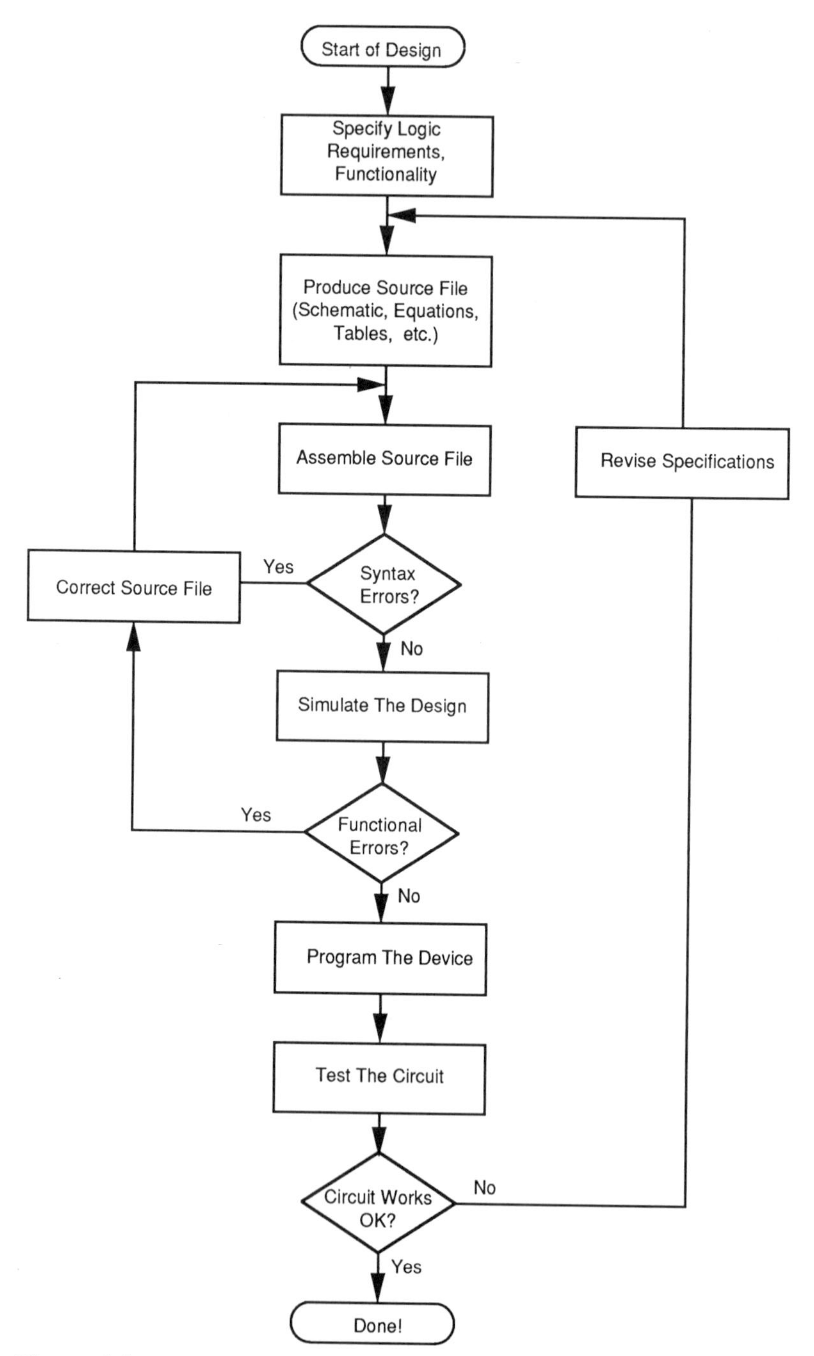

Figure 1-1
Programmable Logic Development Cycle.

Next, a *source file* is produced. Usually, this will be an ASCII source file containing either the equations for PLDs or the program for state machines or a microcontroller. Other possibilities for the source file include schematic diagrams entered on CAD systems, state tables, or waveforms. In all cases, the source file requires careful attention to assure that it is up-to-date, carefully backed up, and accurate.

The source file must be translated into an *object file*. Object files contain the actual bit pattern that will be used to program the device. The object file can be produced by a variety of software. The software is often simply called development software, device assembler, or device compiler. Often the software is provided by the manufacturer of the PLD. One of the exciting developments in the field of programmable logic over the last few years has been the increasing quality and availability of third-party software. Typical examples of vendor-supplied assemblers include PALASM™, AMAZE™, and PLANII™. Third-party software includes ABEL by Data I/O, Inc. and PLDesigner by Minc, Inc. These programs, and other software tools, are discussed in Chapter 5.

For state machines, a meta-assembler is the generic term for the program used to produce the object file from the source file. Typical examples include META29M™ from Paragon Products. Most of these products are targeted at developing a special type of state machine code known as microcode. This topic is covered in greater detail in Chapters 4 and 8.

Microcontroller object files are produced by either assemblers or compilers from ASCII source files. Assembly language maintains an essentially one-to-one correspondence between the mnemonics (the instructions in the source file) and the machines instructions (the code in the object file). Since each type of microcontroller has different machine instructions, assembly language programs are rarely portable to other types of microcontrollers. Compilers allow object code to be generated from high-level languages like C or FORTH. This topic is covered in greater detail in Chapter 11.

The translation of the source (in whatever format) into the object code is a major step in realizing a design with programmable logic. It is usually at this translation step that a variety of format and syntax errors are detected. These include things like conflicting names for symbols or pins, undefined addresses, improper use of output pins, etc. When errors occur (which they almost always do on the first pass), the source file must be corrected before the translation process can continue.

Some software can then simulate the performance of the device. This is very useful for detecting faults in the design. Faults will generally fall into two categories: design rule check (DRC) violations and functional errors. DRC violations include things such as trying to drive a pin to opposite logic levels at the same time. Functional errors occur when the device is behaving the way you *told* it to but not the way you *wanted* it to. In other words, you told it to AND *A* and *B*, but what you really need is for it to OR *A* and *B*. Good simulation software can save long hours of debug time. In some cases, it can catch errors that are almost impossible to capture on the bench.

Once a translation and simulation have been completed, the object code must be loaded into the device. The actual loading is done by a unit called a device programmer. Programmers are typically connected to the computer via a serial channel. Almost always, this is an RS-232 asynchronous channel. A few of the device programmers available plug directly into the expansion slots on PCs; others use the PC's parallel printer port.

In the case of PLDs, the object code is generally used to blow special fuses in the device. This process of blowing the fuses in the device is often referred to as "burning" the device. The nomenclature carries over to other types of devices, and one often hears of burning an erasable programmable read only memory (EPROM) or even an electrically erasable programmable read only memory (EEPROM). This technique of programming by actually burning out fuses is diminishing. More PLDs are going to an EPROM, EEPROM, or even random access memory (RAM) based programming.

State machines are programmed by having the code burned into PROMs or, more often nowadays, EPROMs. The EPROMs are then physically inserted into the circuit. Alternately, special ROM emulator circuits can be used. These emulators are covered in Chapter 12. Simple state machines can be formed from special types of PLDs designed for the purpose. These PLDs are, in general, programmed in the same way as combinatorial PLDs.

Several options exist for linking a microcontroller and its code. Many microcontrollers are programmed with a masked ROM as part of their manufacture. The code pattern is actually built into the part. Needless to say, this does not make changing the code a very viable option. Virtually all microcontrollers today can be purchased with EPROM style memories that can be burned many times. These are very useful during the development phase and are often used for limited production runs. These devices are programmed in a fashion similar to conventional EPROMs.

For system development, two options exist for getting the code to the microcontroller. First, the program may be loaded into extra RAM space and executed by a special program called a monitor. For this technique to work the microcontroller must be linked to the development station. This is usually accomplished via an RS-232 connection. The second and more sophisticated approach is to use an in-circuit emulation vehicle. These topics are developed further in Chapter 12.

However the programmable logic function is implemented, it must be tested to insure that it actually works in the intended application. If everything works, we are home! If the circuit does not work, the "debug" process begins. Assuming that we did the initial simulation correctly, and that the programming went OK, the problem will be isolated to the interaction of the programmable logic block with the rest of the system. The specifications are revisited and adjusted as required. The design cycle then goes through another pass. This process is repeated until a working system is achieved.

1.3 Some Special Concerns with Programmable Logic

Not everything about programmable logic is necessarily a benefit. Along with the many assets come a few liabilities. None of these are particularly serious but if certain precautions are not kept in mind the difficulties can rapidly get out of hand.

The first problem that comes up seems trivial but can actually lead to a lot of confusion. The question is: how can we document the use of a PLD on a schematic? For example, consider the technician who is trying to isolate a problem in the circuit. He is studying the schematic and comes to a box marked "U39." All the inputs are labeled "IN1," "IN2," etc. The outputs are equally informative: "OUT1," "OUT2," etc. This can bring an otherwise productive investigation to an immediate halt. It is generally much clearer to give descriptive names to the input and output pins in the symbol: "ADDR0," "ADDR1," "ENABLE," etc.

Two approaches to defining symbols are shown in Figure 1-2. Figure 1-2a is simply a generic PAL™ and Figure 1-2b is a custom symbol for a programmed PAL™. It should be noted that while Figure 1-2b is the more

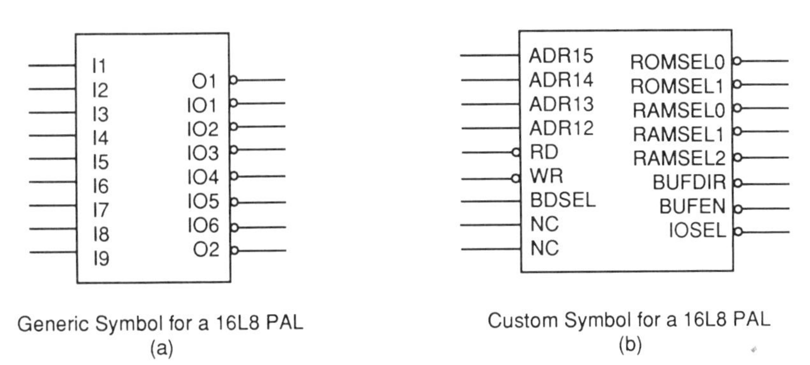

Figure 1-2
Two PAL Symbol Conventions.

descriptive, it is not automatically the better approach. There are often several programmable devices used in a design. The custom symbol approach means that a new symbol must be generated and maintained in the CAD data base for each unique occurrence of the PLD, even if it is the same part except for its programming. Particularly in the early phases of a design, when things may be changing rapidly, this effort could be a waste of time. After all, one of the main tenets of programmable logic is that it allows generic parts to serve custom roles.

Obviously, there is no "right" or "wrong" approach. It depends upon the situation to determine which is the better approach. In the early development phase of a program, it is probably best to stick with a generic symbol. As the design solidifies, a custom symbol can be created for each occurrence of a unique PLD application.

The generic versus custom dual nature of programmable devices extends to other arenas as well. Assigning part numbers is a good example. One is naturally inclined at first thought to simply assign a single part number for a PLD. This soon leads to confusion. When a purchasing department is ordering the part, it might be OK. What about when a customer is ordering the part, however? Does the customer want the "blank" part or the "programmed" part? How does the stockroom even know which is which?

Generally, the best approach is to deal with a PLD as a subassembly. The subassembly bill of material should include the generic unprogrammed part number, the program used to program the part, and a simple description of the part. Ideally, the procedure for burning the part should be included in the subassembly package as well. This subassembly can then be given a unique part number that makes ordering and tracking easier. Such subassembly packages are often called altered item drawings.

The importance of this procedure should not be underestimated. Many hours and considerable sums of money have been wasted when a poorly documented PLD needed upgrading or replacement.

Another criticism of PLDs is that they tend to engender a sense of laziness in designers. It is a complaint, particularly among more experienced designers, that engineers using PLDs are too quick to make use of them; a PLD is invoked whenever a standard off-the-shelf part is not readily at hand. This complaint is often valid. While a PLD can be a real cost saver if it replaces a dozen SSI and MSI chips, it is no real advantage if it replaces only two simple SSI parts.

A frequently overlooked aspect of programmable logic relates to the programming. The use of PLDs can often accelerate the hardware design considerably. This gain must be balanced against the increased time required to program the devices. The trade is not always a favorable one.

Learning curves for the programmable logic devices, availability of programming equipment, design and simulation time, and scheduling of required development equipment must all be factored in.

As we mentioned earlier, these problems are not necessarily reasons to avoid using programmable logic. They are simply examples of the kind of things the experienced designer must take into account when choosing to use programmable logic devices.

Overall, even acknowledging the possible pitfalls in using programmable logic, most designers report that designs are accomplished faster and more economically when programmable devices are baselined.

1.4 Getting Started

As with many activities, getting started in designing with programmable logic is often the hard part. First, it is necessary to obtain at least a minimum hardware configuration to support initial work. This will generally mean a PC of at least an XT level of sophistication. If sophisticated third-party software is going to be used, an AT with EGA or VGA monitor is recommended. A printer is mandatory but usually the type is not particularly important. Simple dot matrix types will generally work fine. Hard disc storage requirements are often reasonable. A 20 Mbyte drive or larger is generally adequate. If a computer is not already available, the setup described will cost an estimated $2,500 (U.S.).

Software for generating the object files is required. For simple designs, the manufacturer or distributor will often provide software packages for free. Many of these packages are of a surprisingly high quality. For large projects, however, third-party software is often a good investment. Third-party software will cost $1,500 to $5,000.

Access to a device programmer is required. For limited use, or on a trial basis, many distributors will make their in-house units available to customers. Alternatively, relatively inexpensive device programmers may be available. Simple units will cost anywhere from $500 to $2,500. More sophisticated units range from $5,000 to $10,000. Plan on the higher number for serious work with PLDs. A few microcontrollers are designed to be self-programming. Motorola has several parts in this category. For these chips, programmers may cost as little as $150.

Next, one should choose a design that is sufficiently challenging to exercise the selected devices but simple enough to be unambiguous. In using the process shown in Figure 1-1, try several different ways of realizing the design. This will develop experience and confidence as well as point out any weak links in the development cycle.

Programmable logic design is a field where "creative play" often pays off. In the hands of an experienced designer, programmable logic techniques can often yield amazing results in very short periods of time.

On the other hand, it is extremely frustrating to be learning new architectures and procedures when critical deadlines are fast approaching. Devices are often subtly different than what one would anticipate from reading their data sheets. New discoveries are best appreciated when time is not short.

This suggests that engineers, particularly new engineers, be allowed an unpressured learning period when first working with programmable logic. An unpressured period of learning and experimentation is important for several reasons. It allows one to catch and correct points of confusion off-line from the actual design, thus making it easier to openly acknowledge mistakes and ask for assistance and advice. Furthermore, taking the time to learn and experiment with new devices and techniques often pays dividends. Opportunities for cost savings and design simplification are more likely to be spotted and capitalized upon. It is often far more productive to simply allocate a learning and experimentation period at the start of a design than it is to have a design not function due to a designer's misconception.

The expense in developing programmable logic designs generally breaks down into three major categories: personnel time; equipment setup costs; and the cost of the programmable devices. Generally, the magnitude of the costs will be in the same order, with personnel time being the major cost. The cost of the devices themselves will generally be trivial in comparison to the other costs.

This cost profile indicates that getting people up to speed rapidly and executing the design process efficiently are the main goals in effectively developing programmable logic designs. Money spent on good tools will generally be well invested. Providing for sample parts that can be used to test concepts and try alternatives will also provide a good return on investment. These cost minimization strategies are often lost on management personnel if they have not "been in the trenches" themselves.

Programmable logic provides a flexible, cost-effective, and interesting approach to solving many design problems. The rest of this book is devoted to helping the reader get the most out of the various programmable logic techniques.

1.5 Chapter Summary

- Programmable logic design is the art of applying software techniques to hardware problems.

- Programmable logic trades hardware design time for software design time and usually the trade is a favorable one. This is especially true if flexibility in the design is required or if ambiguity in the design exists.
- Learning curves should be taken into account when scheduling designs. Generally, two days of learning time is enough for simple PALs or field programmable logic arrays (FPLAs). Several weeks should be allocated for sophisticated "programmable gate arrays" and microcontrollers.
- If programmable logic development stations and programming equipment are not already in place, allow generous time to get them set up. Installing hardware, software, cables, and procedures can take a week or more.
- Using programmable logic allows a single chip to be used for different purposes in different parts of the design. This lowers the cost of procuring and maintaining inventories.
- While programmable logic design simplifies inventories, care must be taken in the documentation and programming of the individual parts. Failure to do so can lead to confusion and lost time, resulting in increased expenses.

2

Basic Logic Design

This chapter is a brief review of the fundamentals of logic design. It is divided into two sections. The first is an overview of the design process and high-level definitions of the topics to be covered in later chapters.

The second part is a review of basic logic functions. The focus is on the elementary aspects of logic that are the most useful for understanding and using programmable logic. From a practical perspective, much of the mathematical manipulation done in undergraduate courses in logic is of little real use to the working engineer. Most software packages, and all of the good software packages, will perform logic reduction and minimization for you. Furthermore, the optimization of the logic function is often device dependent. Again, the better software packages take this into account and free the engineer from the burdensome task of making the logic function fit the architecture.

Even with the sophistication of modern software tools, however, some of the mathematical techniques are worth reviewing. There are two reasons for this. First, it is always useful to have a solid understanding of a problem before turning it over to a computer. Even if the computer can do a better job, we are at least in a position to double check the machine's work. Debugging, change evaluation, and circuit modification are all easier if we have a good understanding of the basic logic.

Second, and most importantly, the architectures of PLDs are directly dependent on the theoretical concepts presented here. Single-plane and dual-plane logic architectures can only be understood by appreciating the fundamental logic that drives their architecture. De Morgan's theorem and canonical forms are of more than academic interest; they directly determine what functions can be accomplished by which devices.

Engineers often feel that they have used PLDs successfully in designs but they did not really understand the devices or the processes involved. This lack of understanding often follows from not having the theoretical basis for the PLD's architecture clearly presented. Later chapters will relate the theoretical concepts reviewed in this chapter to practical device architectures; this will make the "whys and wherefores" of PLDs more easily understood.

2.1 Logic Design

Logic design can be viewed, if we take a sufficiently abstract position, as consisting of two basic tasks. The first is to define some set of Boolean outputs for a given set of Boolean inputs. Hopefully, as part of this definition, we can find some function f that relates the two. As we will see shortly, however, it is not always necessary to figure out what f actually is.

The second task is to realize (sometimes called synthesize) a circuit to yield the output, y, for any given input, x. An overview of this perspective is shown in Figure 2-1. The values x_1, x_2, x_3, etc. are a sequence of binary numbers that we input to our circuit. For each input x_i there is an output y_i. For example x_1 yields y_1, x_2 yields y_2, x_3 yields y_3, etc. These y values are the circuit's responses, defined by the function f, to the inputs x.

If each y_i is completely defined by each x_i only, then f can be realized by a *combinatorial circuit*. If y_i depends on both the x_i and some previous value of x_{i-j}, then f must be realized with some form of *sequential circuit*, generally called a *state machine*.

When we decide to realize the circuit, there are two basic approaches we can take. The first is to build a circuit to compute f. This is what we do when we design a conventional circuit using conventional components such as AND gates, OR gates, INVERTERs, etc. These conventional circuit design techniques are often called discrete or hardwired techniques.

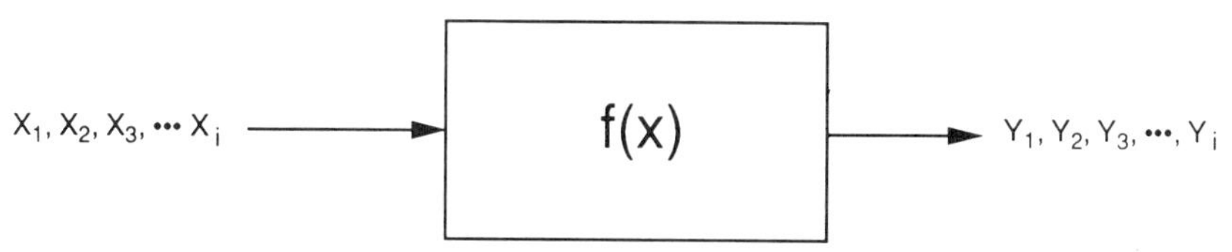

Figure 2-1
Abstract View of a Digital Circuit.

The second approach is to program some engine to compute *f* for us. This approach is the subject of this book. We will look at conventional design techniques only to the extent of comparing and contrasting them with programmable design techniques.

With this idea of programming a circuit to compute *f*, we offer the following definition: Programmable logic design is the realization of a circuit for computing some Boolean function by employing a programmable engine. In other words, we wish to implement a logical circuit that will exactly model the circuit we would get if we were to build a discrete version of the circuit.

The advantages of programmable logic are that we do not have to build a discrete circuit with all of its individual wires and components; the programmable circuit is far more flexible and changeable; and the same type of programmable device can be used for many different functions. Additionally, a circuit realized with programmable logic devices will generally require fewer devices and less physical area. These advantages lead directly to more efficient, reliable, and cost-effective designs.

We can realize our programmable engine in one of three basic ways:

1. We can make physical or logical connections in a specially structured logic array that has been designed for programming. By doing this we are effectively modeling a hard-wired design that will compute *f*. An example of this is the use of a programmable logic devices (PLDs) such as programmable array logic (PAL), or programmable logic arrays (PLAs). We call this technique *emulating a discrete logic design.*
2. We can simply tabulate the logic output for each possible combination of inputs. The resulting table is stored in a memory device, typically a PROM. This is the technique often used in constructing a state machine and occasionally in implementing combinatorial circuits. Note that there are two interesting properties of this technique: First, we do not need to know the function we are implementing. As long as we know what output we want for which input, we can realize our goal. Second, the overall architecture of this approach lends itself very well to sequential operations. Examples include Mealy state machines, Moore state machines, and microsequencers.
3. We can use a general purpose Von Neuman machine to compute, using an *algorithm,* the desired output for any given input. This is what we do when we make use of a *microcontroller*. Advantages of microcontrollers are the very complex algorithms that can be implemented, the ease of programming the algorithms, and the sophisti-

cated peripheral support functions such as timers and I/O available on microcontrollers.

The relative merits of these three techniques are shown in Figure 2-2. Note that PALs, PLAs, and PROMs, being simple combinatorial circuits, are the fastest. For this reason they are often used to implement address decoders and other speed sensitive circuits. State machines are not as fast as simple PLDs. On the other hand, they can handle *sequences* of operations. They are not as flexible as microcontrollers but often they are orders of magnitude faster. For this reason state machines are often used in bus arbitration circuits, waveform generators, high-speed sequencers, and other fast sequential applications.

If we disregard speed, microcontrollers span the range of applications for programmable logic. Theoretically, anything that can be done by a PLD or a state machine can be programmed into a microcontroller. One might assume that there would be a cost penalty (in terms of dollars) associated with using a microcontroller in simple applications but this is not necessarily so. Microcontrollers are produced in such quantities that they can often be cost-effective even for simple applications. This point is emphasized because many engineers, oriented toward thinking only in hardware terms, choose PLD or state machine designs when a microcontroller is a simpler, more cost-effective solution. It is easy to assume

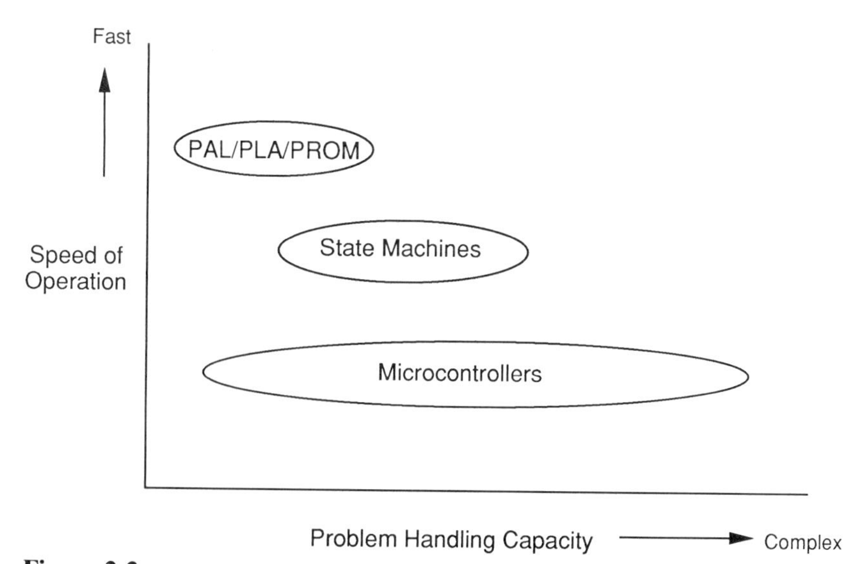

Figure 2-2
Programmable Logic Options.

(incorrectly) that a microcontroller's increased sophistication will automatically make it more expensive and harder to use.

2.2 Selected Topics in Basic Logic

This section covers Boolean logic, rules of precedence, and basic logic operation. The use of De Morgan's theorem is touched upon. Latches and flip-flops play an important part in many designs using programmable logic. The RS and D type latches are presented. Our discussion of flip-flops is limited to the ubiquitous J-K type. A good understanding of latches and flip-flops is useful when we look at "macrocells" in Chapter 5.

The final part of this chapter presents a simple circuit: a push button operated counter. We use this counter function several times in the book to make comparisons between various logic implementations.

Before starting, some useful definitions are in order. Discussions of logic, like many other technical discussions, can be hard for the uninitiated to follow. It is difficult to know when two terms are synonymous and when different terms really have a different meaning. For example, saying that "The active high signal was a one" is synonymous with saying that "The positive logic signal was high."

These terms can best be understood by starting with the basics and working our way up. Here are some brief and nonrigorous definitions and conventions:

- While it is not theoretically necessary, in practice all common digital logic is based on binary arithmetic. Binary numbers have only two digits: 0 and 1.
- These values are represented in a logic circuit with voltage potentials that are above or below a certain *threshold voltage*.
- By definition, the "threshold" voltage is the level at which a signal changes from a zero to a one, or a one to a zero. In some forms of logic, like TTL, the threshold voltage is not the same for a "one to a zero" transition and "zero to one" transition. There is a small "dead zone" in between the transitions, where the state of the signal is not defined.
- A binary 1 is, by convention, a "high" (i.e., above the threshold) voltage. A binary 0 is a "low" voltage.
- In classic Boolean algebras the terms 0 and 1 are not used. Boolean algebras are concerned with *logical operations,* not arithmetic ones. For Boolean logic, the values of interest are TRUE and FALSE.

Like binary values, the TRUE value is conventionally represented by a high, the FALSE value by a low.

The terms 0 and 1 and the terms TRUE and FALSE are sufficiently close in meaning that they are quite often interchanged. Truth tables, for example, are often drawn with 0s and 1s. The practice of interchanging the meanings between binary and logic values does no real harm. The translation is done by most experienced engineers without conscious thought. The object of this discussion is not to make a stand on semantics. It is important, however, to realize that there are differences. This can be quite confusing to those who are new to the working world of digital design.

A third consideration comes into play with respect to digital *circuitry*. The output of a logic circuit, for example, may be an LED indicator. In this case, the meaning of 0 or 1, or TRUE and FALSE, is not clearly discernable. What is important is: Under which conditions is the LED producing light? The LED is said to be active if it is producing light and inactive if it is not.

Now, an LED can be wired to the output of a logic gate so that current flows when the gate is a high, or an LED can be wired so that current flows when the output of the gate is a low. These two cases are shown in Figure 2-3. In Figure 2-3a the LED will be *active when the gate is low*. In Figure 2-3b the LED will be *active when the gate is high*. These expressions are generally abbreviated to *active low* and *active high*. Notice that, depending upon the configuration, either a 0 or a 1 can produce the active state.

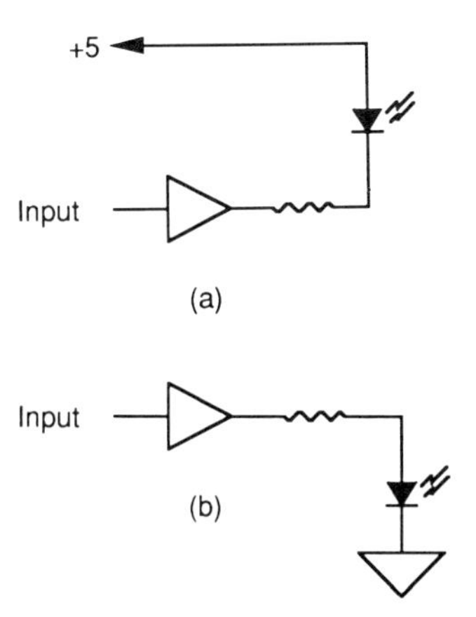

Figure 2-3
Active Low versus Active High Circuits.

Even the most experienced designers can get confused from time to time. When there are too many "lows," "ones," and "active this," and "active that" floating around in the conversation, it is worthwhile to take the time to stop and make sure that everyone is on the same wavelength.

Before proceeding to the discussion of logic, it is important to define some other conventions found in the literature. The following are some of the more standard conventions:

- The AND operation is often indicated by a dot, as in the dot product symbol. Alternately, ANDing is often implied by simply writing two variables together: AB means A AND B. It is from this convention that ANDing two variables is often called "taking the product of the variables."
- The OR operation is often shown by use of the "+" sign. $A + B$ means A OR B. Again, from this convention OR operations are sometimes referred to as "summations."
- Exclusive OR (XOR) operations are shown as a "+" symbol with a circle around it, as "$\oplus$" (as in $A \oplus B$), or as a V with an underline.
- The unary NOT operation is often shown as either a line over the variable or as an apostrophe after the variable: NOT $A = A'$. Usually in textual contexts, particularly with programmable logic, the NOT operation is shown as a "/" sign: NOT A = /A. Occasionally, the C programming language convention is adopted: NOT A = !A.

These conventions are found throughout the literature and are mixed between text and illustrations fairly regularly. This is confusing at first and has the effect of making things look more complicated than they really are. Once one gets used to the different ways things are expressed, it becomes automatic to make the translations in one's head.

In this book the following conventions are adopted:

- AND operations are indicated with the "*" character or explicitly called out as "A AND B."
- A bar is placed over a variable to indicate the NOT operation. In programming examples, or other "text-oriented" discussions, either a "/" is used in front of the variable to indicate the NOT operation or the C operator "!" is placed in front of the variable.
- Exclusive OR operations are shown as a "$\oplus$" or explicitly called out as "XOR."

2.2.1 Combinatorial Logic

Combinatorial logic, as the name implies, is composed of logic elements that act in various combinations to produce a logically useful output. Tabulating and manipulating the possible combinations is a large part of logic design.

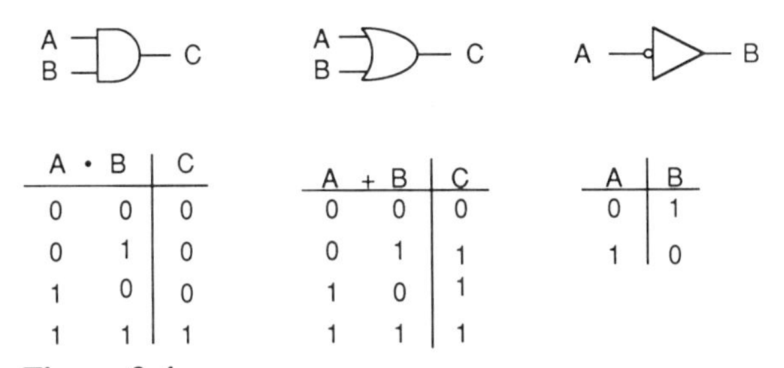

Figure 2-4
Basic Logic Gates.

By definition, combinatorial logic is limited to acting on the immediate inputs to the system. Combinatorial logic makes no use of, and is independent of, any previous inputs. In Section 2.3 this constraint is removed when sequential circuits are reviewed.

Basic Gates Figure 2-4 shows the three standard logic gates that form the basis of most digital design. These gates are the AND gate, the OR gate, and the NOT gate (or INVERTER). From these basic gates, any Boolean (or binary) function can be built up. Mathematically, these functions are simply represented as: $A * B$ for the AND gate; $A + B$ for the OR gate; and simply $/A$ for the inverter. Often, the full symbol for an inverter is not used. The circle is simply attached to the input or the output of a gate to indicate that the signal is being inverted. Examples of this are the Not-AND (NAND) gate and the Not-OR (NOR) gate.

Figure 2-5a shows a more complicated gate called an Exclusive OR (XOR) gate. The XOR gate has the interesting property of being TRUE when one and only one input is TRUE. This is a good example of how

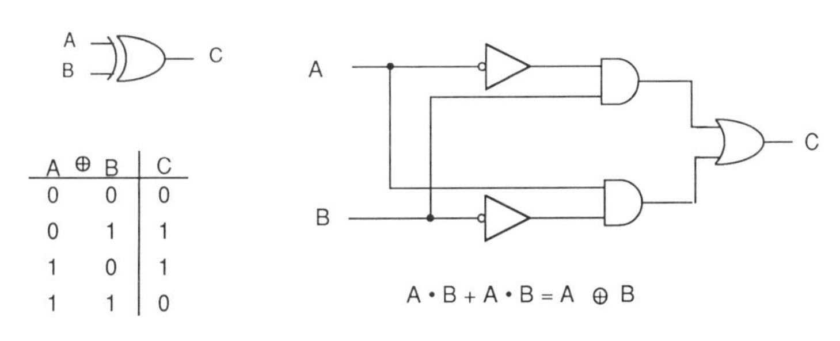

The XOR Gate and Its Truth Table

(a)

XOR Gate Built from Simple Gates

(b)

Figure 2-5
The XOR Gate.

more complex functions are built up from standard gates. Mathematically, the XOR function is defined as:

$$(A * \overline{B}) + (\overline{A} * B)$$

Figure 2-5b shows a possible logic implementation of this function. XOR gates find use in binary mathematics circuits such as adders and subtractors.

An interesting use of the XOR gate is as a programmable inverter. For example, most microcontrollers have interrupt lines that are sensitive to active low logic levels (this allows for WIRE-ORing multiple inputs). The interrupt source, however, may or may not be an active low signal. Figure 2-6 presents a simple and universal way of resolving this problem: a programmable inverter. If the switch is open, the XOR gate acts as an inverter. If the switch is closed, the circuit simply passes the input to the output unchanged.

XOR gates play an important role in some of the more flexible PLD architectures. The "programmable inverter" feature allows efficient utilization of both active low and active high logic functions.

The notion of an "active low signal" introduces the concept of positive and negative logic. Positive logic is simply what we would normally think of when looking at the sense of a logic gate. A logical 1 is some positive voltage, typically positive 5 V. Negative logic is what we get when we want a signal to be TRUE (i.e., active) when it is low or close to ground potential. To illustrate this concept, let us look at a typical indicator circuit, such as the one shown in Figure 2-7. We only want the LED to glow when both input *A* and input *B* are low.

Figure 2-7 shows two ways of approaching the problem. Figure 2-7a is a classic "positive logic" implementation. This is derived from simply following the requirements of the design. First, the two sources are inverted, so that their signals can be logically ANDed together; second, the signals are ANDed; and third, the result of the ANDing is INVERTED to make it active low. Mathematically this can be written as:

$$\overline{I} = \overline{A} * \overline{B}$$

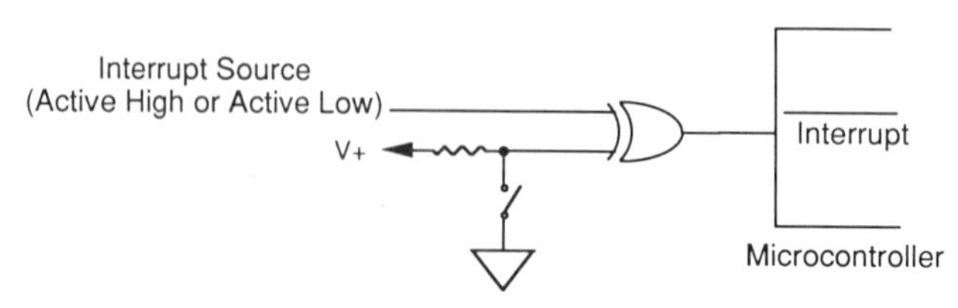

Figure 2-6
XOR Gate Used as a Programmable Inverter.

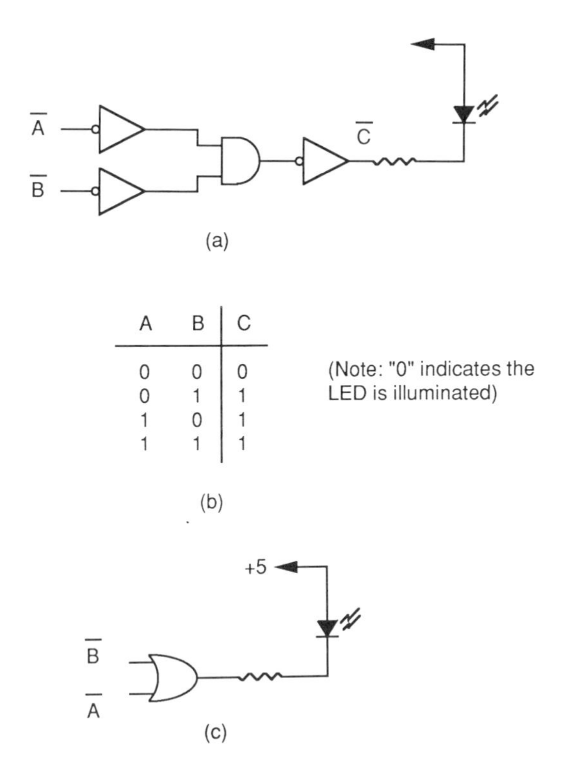

A	B	C
0	0	0
0	1	1
1	0	1
1	1	1

Figure 2-7
Examples of Positive and Negative Logic.

On first examination, this might seem as simple as we can make the circuit. If we examine the truth table for this circuit (Figure 2-7b), however, we note that it matches the truth table for a simple OR gate. Thus we can replace three inverters and one AND gate with a simple OR gate. This identity holds in the math as well:

$$\overline{I} = (\overline{A} * \overline{B}) = \overline{A + B} \tag{2-1}$$

In fact, this relationship is formally known as De Morgan's theorem. In words, it says that a positive logic AND gate (or function) is identical to a negative logic OR gate (or function). Thus Figure 2-7c can be used to implement a circuit functionally identical to the circuit shown in Figure 2-7a.

The relative cost of the two forms should be noted. The positive logic version of our circuit required three inversions and an AND operation. The negative logic version required only the one operation. Exploiting this technique efficiently can mean substantial parts savings when building real world hardware.

Mathematical Manipulation of Logic Equations Mathematical manipulation of logic functions is often necessary. Sometimes the manipulation is required to adjust the form of a logic equation to the architecture of a PLD. Sometimes the manipulation is necessary to simply get a good feeling of how the design is behaving.

The rules for mathematically manipulating logic functions are tabulated in Figure 2-8. These rules can be used to either expand or simplify a logic function, as required. In general, manipulating logic equations follows the sense of manipulating equations algebraically. As demonstrated by De Morgan's theorem, however, there are traps for the unwary. The NOT operator distributes in the same way the negation operator distrib-

Properties of Logic Operations:

A • (B • C) = (A • B) • C	AND is associative
A + (B + C) = (A + B) + C	OR is associative
A • B = B • A	AND is commutative
A + B = B + A	OR is commutative
A • (B + C) = (A • B) + (A • C)	AND is distributive over OR
A + (B • C) = (A + B) • (A +C)	OR is distributive over AND

Simplification Theorems:

A • A = A

A + A = A

A • 1 = A

A + 1 = 1

A • 0 = 0

A + 0 = A

A • /A = 0

A + /A = 1

A • (A + B) = A

A + A • B = A

A • (/A + B) = A • B

A + /A • B = A + B

//A = A

De Morgan's Theorems:

/(A • B) = /A + /B

/(A + B) = /A • /B

Figure 2-8
Properties of Logic Operations.

utes over variables but the NOT operator does *not* distribute across logical operations. For example:

$$\overline{(A * B)} \neq \overline{A} * \overline{B} \tag{2-2}$$

The nice thing about logic equations is that they are easy to compute since the variables always have a numerical value of either 0 or 1. This makes it a simple matter to check our math as we go.

One of the key uses of manipulating logic equations is the simplification of an equation. "Simplification" is one of those things that is easy to hang a label on but cannot be universally defined. The "simplest" implementation of a logic function depends on the environment. A three- or four-level logic network may be simplest, in the sense of having the fewest terms. On the other hand, such a network can always be simplified down to a two-level network. Usually, decreasing the number of levels comes at the cost of increasing the number of terms or the size of the terms in the equation.

In general, a simplified logic equation follows the rules associated with a simplified algebraic equation: the smallest number of expressions (common factors eliminated, parentheses eliminated, etc.).

Two other terms that are often associated with simplified logic are minimization and optimization. Minimization specifically refers to reducing to a minimum the number of product terms required to implement a logic function. This can also be thought of as requiring the fewest resources of the implementation vehicle to realize the given function.

Optimization refers to realizing the design in the most optimum fashion. The optimum solution is not necessarily the simplest or minimalist solution. Optimization criteria have to be stated for the term to have much meaning. Common optimization criteria include speed, routing density, pin utilization, etc.

Simplification, optimization, and minimization are sometimes used interchangeably. This is a practice to be discouraged since the terms each have a unique meaning. The misuse of the terms occurs often enough, however, that it is worth checking to see how the terms are being used if you are unsure.

To illustrate these concepts, let us look at an example:

$$F = (\overline{A} * \overline{B} * \overline{C}) + (A * \overline{B} * \overline{C}) + (\overline{A} * B * \overline{C}) \tag{2-3}$$

The variable /C is common to all three terms. It can therefore be factored out:

$$F = \overline{C} * (\overline{A} * \overline{B} + \overline{A} * B + A * \overline{B}) \tag{2-4}$$

This is further factored as:

$$F = \overline{C} * (\overline{A} * (\overline{B} + B) + A * \overline{B}) \tag{2-5}$$

The expression ($/B + B$) is always true, so it can be dropped from the equation, which yields:

$$F = \overline{C} * (\overline{A} + A * \overline{B}) \tag{2-6}$$

Since ($/A + A/B$) is identical to ($/A + /B$), Equation 2-6 can be further simplified to:

$$F = \overline{C} * \overline{A} + \overline{B} \tag{2-7}$$

We multiply everything out to eliminate the parentheses. This leaves us with:

$$F = \overline{A} * \overline{C} + \overline{B} * \overline{C} \tag{2-8}$$

Equation (2-8) is a simplified version of Eq. (2-3) since it is written as simply as the function can be expressed. It is minimized since it has the fewest product terms needed to realize the function. Whether or not it is an optimal form of Eq. (2-2), however, depends upon what resources are available for realizing the function.

Canonical Forms As noted previously, tabulating logic functions, at least relatively simple ones, is easy because of the binary nature of the variables. The tabulation process is important since it allows us to get the feel for whether a particular logic function can be fit into a particular architecture.

Consider an arbitrary logic function of three variables: A, B, and C. What is the worst case number of input conditions we must deal with? The answer can be seen by examining Figure 2-9. Every possible input combination is listed. By counting the number of rows, we find the maximum number of input conditions. In this case the answer is eight. By doing this for different numbers of variables, we rapidly discover that the possible number of input conditions is simply 2^n.

Equation (2-8) is said to be a sum of products (SOP) form. The product terms, $/A/B/C$ for example, are also called the minterms of the expressions. Under the column marked ''minterms'' in Figure 2-9, there is a number for each row. This number can be used to designate the terms that we use in a logic equation. For example:

$$F = (\overline{A} * \overline{B} * \overline{C}) + (\overline{A} * B * \overline{C}) + (A * \overline{B} * \overline{C}) \tag{2-9}$$

can also be written as:

$$F = m_0 + m_2 + m_4 \tag{2-10}$$

When every variable (A, B, or C) participates in every term, the logic function is said to be expressed in its *canonical form*. We can see that, at most, we will need eight *product* terms summed to completely

Address	Minterm		Maxterm	
000	m_0	$\bar{A}\bar{B}\bar{C}$	M_0	$A+B+C$
001	m_1	$\bar{A}\bar{B}C$	M_1	$A+B+\bar{C}$
010	m_2	$\bar{A}B\bar{C}$	M_2	$A+\bar{B}+C$
011	m_3	$\bar{A}BC$	M_3	$A+\bar{B}+\bar{C}$
100	m_4	$A\bar{B}\bar{C}$	M_4	$\bar{A}+B+C$
101	m_5	$A\bar{B}C$	M_5	$\bar{A}+B+\bar{C}$
110	m_6	$AB\bar{C}$	M_6	$\bar{A}+\bar{B}+C$
111	m_7	ABC	M_7	$\bar{A}+\bar{B}+\bar{C}$

Figure 2-9
Minterms and Maxterms of Three Variables.

define any arbitrary logic function of three variables. As we will see in future sections, these estimates can be important in choosing a suitable logic device.

Figure 2-9 lists both the minterms and the maxterms possible for a function of three variables. These are important since they provide a vehicle for changing the form of the equation. Changing the form from a SOP to a product of sums (POS) form is straight forward. F can be expressed equivalently by simply taking the product of the maxterms and complementing:

$$F = \overline{(A + B + C) * (A + \overline{B} + C) * (\overline{A} + B + \overline{C})} \qquad (2\text{-}11)$$

$$F = \overline{M_0 * M_2 * M_4} \qquad (2\text{-}12)$$

The reason that this transformation is important is that it allows the use of different hardware to implement the same function. This is handy when, for example, only the AND gates in a device are programmable. These and other architectural issues will be presented in Chapter 3.

In general, any combinatorial logic function can be expressed as either a SOP or a POS. The general form for a SOP is:

$$F = a_0 m_0 + a_1 m_1 + \cdots + a_k m_k = \sum_{i=0}^{k} (a_i m_i) \qquad (2\text{-}13)$$

where m_i is the minterm selected by a_i. If a_i is a 1, the minterm is included in the expansion. If a_i is a 0, the minterm is dropped from the expansion.

A POS function can be expressed in general as:

$$F = (A_0 + M_0)(A_1 + M_1)(A_2 + M_2) \cdots (A_k + M_k) = \prod_{i=0}^{k} (A_i + M_i) \tag{2-14}$$

where M_i is the maxterm selected by a_i. If a_i is a 1, then $(a_i + M_i)$ is always a 1 and can be dropped from the expression.

Canonical forms, SOP, and POS, along with De Morgan's theorem form the very heart of programmable devices such as PLDs, PALs, and ROMs. These devices will be discussed in the next chapter. For the moment, however, let us look at some important aspects of these mathematical relationships.

Figure 2-10 shows a typical combinatorial logic network. Notice that there are three levels of logic involved. The logic equation for this network can be obtained by inspection:

$$F = (A * B + C) * \overline{D} \tag{2-15}$$

It will be important for later discussions to understand how this function can be expressed in canonical form. There are two approaches to make the conversion: The first is to use Figure 2-8 and manipulate the logic equation. This is an interesting exercise but it is somewhat tedious and will not be pursued in this text.

The second approach is to build a truth table such as the one shown in Figure 2-11. All possible inputs and outputs are tabulated. From the table in Figure 2-11 the canonical form of the function is easily expressed:

$$F = m_2 + m_6 + m_{10} + m_{12} + m_{14} \tag{2-16}$$

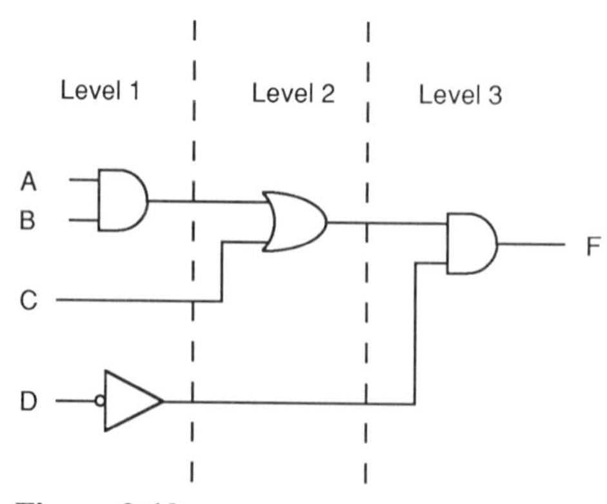

Figure 2-10
Three Level Logic Circuit.

$$F = (A * B + C) * \overline{D}$$

A	B	C	D	F
0	0	0	0	0
0	0	0	1	0
0	0	1	0	1
0	0	1	1	0
0	1	0	0	0
0	1	0	1	0
0	1	1	0	1
0	1	1	1	0
1	0	0	0	0
1	0	0	1	0
1	0	1	0	1
1	0	1	1	0
1	1	0	0	1
1	1	0	1	0
1	1	1	0	1
1	1	1	1	0

Figure 2-11
Truth Table for Figure 2-10.

Notice that the minterms are selected by the presence of a 1 in the output column of the truth table shown in Figure 2-11.

Equation (2-16) can be implemented with a two level network. Such an implementation is shown in Figure 2-12. In comparing Figure 2-10 with Figure 2-12, one may question why we would want to bother with the effort. After all, the circuit in Figure 2-10 uses only three simple gates and an inverter. The circuit in Figure 2-12 requires six moderately complicated gates.

The answer is that the circuit in Figure 2-10 is quite unique to the function *F*. While the circuit in Figure 2-10 is simpler than the one shown in Figure 2-12, it does require that three different types of gates be wired together in a specific way. The circuit in Figure 2-12 is much more regular. Notice that all of the AND gates are connected to the one OR gate. The order in which the AND gates are connected is not important. Also, with the exception of the sense of the inputs, all of the AND gates are the same.

This regularity offers two main advantages. First, the delay time through the circuit is uniform. The circuit in Figure 2-10 can have race conditions in which the one input signal reaches the output before the

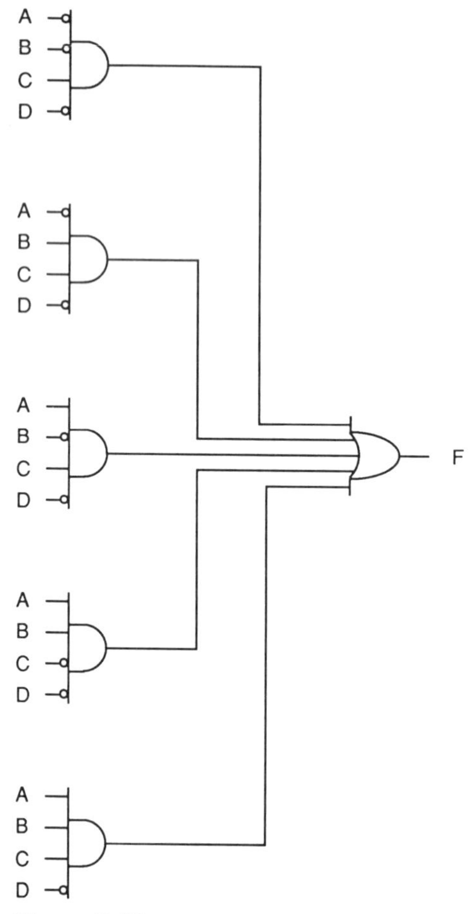

Figure 2-12
Two Level Equivalent of Figure 2-10.

other inputs have had a chance to affect the output. This can be avoided in either circuit but it is easier to accomplish with the circuit in Figure 2-12. Second, the regularity of the circuit in Figure 2-12 lends itself well to automation and therefore to programmable implementations.

A graphical way of reducing logic equations, called Karnaugh mapping, is popular. Karnaugh mapping replaces the algebraic manipulations with geometrical mapping. The technique is limited to two or at most three variables. This limitation is due to the geometrical nature of the mapping. One axis is required for each variable: two variables map into a plane, and three variables map into a cube. Beyond pointing out that it exists, we do not discuss Karnaugh mapping in this book. The reasons are twofold: First, the technique is not used that often in programmable logic design. Second, the technique is well covered in textbooks on logic (see the annotated bibliography for a list of titles).

Other techniques are available for reducing logic equations. The most popular is probably Quine–McCluskey. These techniques are often used by PLD compilers for automated reduction of logic equations. The Quine–McCluskey technique is not well suited for reducing logic equations by hand, however.

To sum up the discussion of programmable logic, we emphasize two main points: First, any function can be realized by simply ORing together selected minterms. (This will be further illustrated when we discuss PROMs in the next chapter.) Second, logic equations can be manipulated as necessary to modify or simplify their function.

2.2.2 State Logic

Many interesting functions can be accomplished with combinatorial logic. Sequential sorts of activities such as counting cannot be realized with purely combinatorial circuits, however. To realize sequential operations it is necessary to feed part of the *output* of a logic network back to the networks *input*. This, in effect, forms a memory that allows the network to know where it is in a sequence.

A large number of ways exist for forming this feedback path. The simplest is to wire several of the outputs directly to some of the inputs. As will be shown below, this is exactly how latches and flip-flops are built.

The feedback path may itself be a clearly definable logic function. In this case, the feedback path consists of programmable logic elements. Techniques for realizing these types of sequential circuits will be shown in upcoming chapters.

To accomplish sequential activities, the concept of *state* must be introduced. The simplest example of state logic is a reset/set (RS) latch. A simple circuit is shown in Figure 2-13a. When the NAND gates are cross-coupled as shown, the circuit has the capability of remembering its input even after the input has been removed. In other words, a momentary logic low on the /SET input will cause the Q output to go high (if it was low) and to stay high. The $/Q$ output is the opposite of the Q level, so it will go low (if it was high) and stay low.

This action forms the basis of the definition of *state*. The latch is said to be in the high-state if the Q output is high. The latch is said to be in the low-state if the Q output is low. Figure 2-13b is the normal schematic symbol for an RS latch.

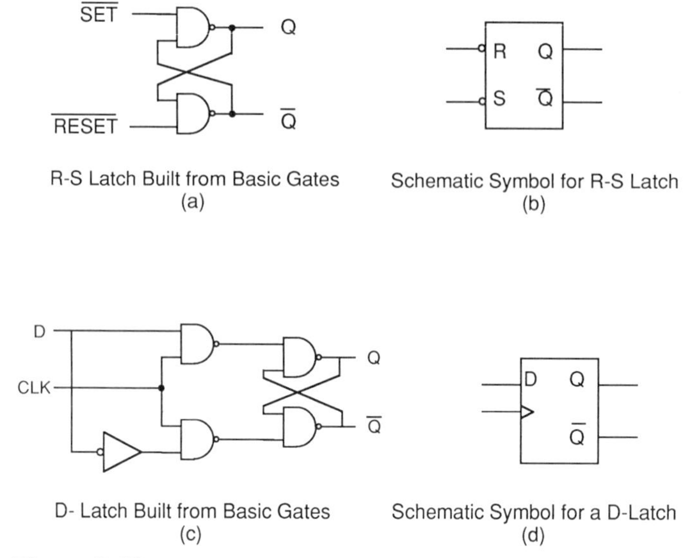

R-S Latch Built from Basic Gates
(a)

Schematic Symbol for R-S Latch
(b)

D- Latch Built from Basic Gates
(c)

Schematic Symbol for a D-Latch
(d)

Figure 2-13
Basic Latches.

The RS latch by itself is not terribly entertaining. Nevertheless, the RS latch does find use in static memory circuits and in-switch debouncing circuits. A more interesting circuit is developed when an RS latch is combined with some simple combinatorial logic. The simplest form of such a circuit is the clocked latch also known as a "D-latch." D-latches are very useful as output ports and for holding an image of the data after data has gone away. A gate level implementation for a D-latch is shown in Figure 2-13c. D-latches are often used to demultiplex combined address/ data busses. The output of the latch follows the data as long as the clock is a 1. When the clock goes low, the output is frozen and any changes on the input of the latch are ignored. Figure 2-13d is the schematic symbol for a D-latch.

An interesting circuit based on the RS latch is shown in Figure 2-14a. This circuit is normally called a master/slave flip-flop because of its architecture. The circuit operates by clocking data into the master RS latch on the high transition of the clock. On the low transition of the clock, the data is transferred to the slave. This two-part action is fundamental to the operation of the flip-flop.

The gating around the master provides for a variety of options on how to use the flip-flop. If both the J and the K inputs are low, the circuit remains in its current state. If the J input is high and the K input is low, the Q output will go high on a clock cycle. If the J input is low and the K

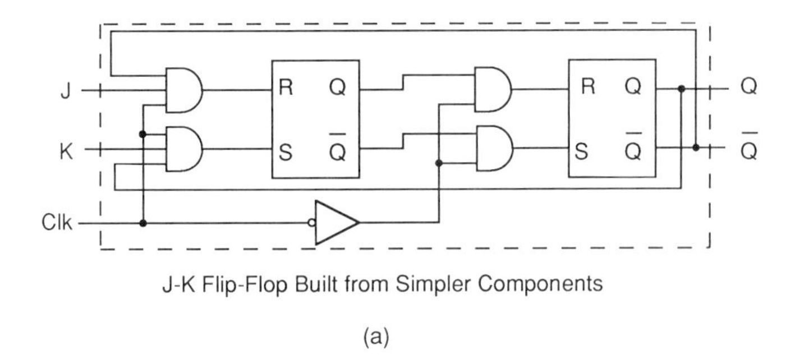

J-K Flip-Flop Built from Simpler Components

(a)

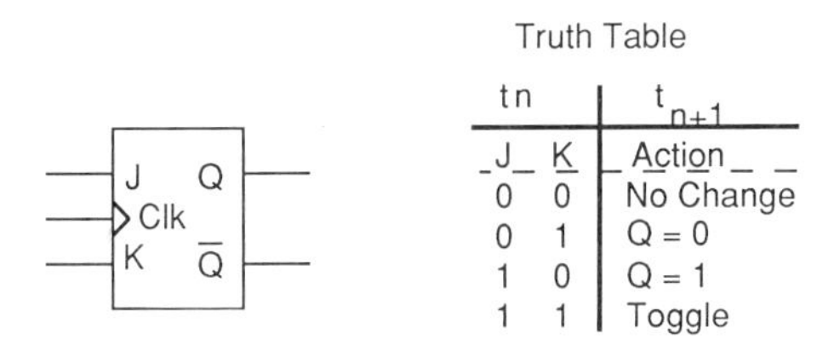

Truth Table

t_n		t_{n+1}
J	K	Action
0	0	No Change
0	1	Q = 0
1	0	Q = 1
1	1	Toggle

J-K Flip-Flop Symbol and Its Truth Table

(b)

Figure 2-14
The J-K Flip-Flop.

input is high, then the Q output will go low on a clock cycle. The final case, and the most interesting case for our immediate discussion, is when both inputs are high. When this occurs, the flip-flop will toggle between states on each clock cycle (from which it gets its name, flip-flop).

This toggling forms the basis from which a counter can be constructed. Figure 2-14b is the symbol for a master/slave flip-flop (also called a J-K flip-flop from the labeling of the two inputs). Figure 2-15a shows a simple 2 bit counter formed from two master/slave flip-flops. Since there are two flip-flops, each of which can assume two states, our circuit is capable of counting to four. The four possible states are 00, 01, 10, and 11. A "state diagram" of this circuit is shown in Figure 2-15b.

While the counter in Figure 2-15 is useful in many applications, it has potential problems. Each stage of the counter will introduce some delay from the input to the output. Thus the clock will ripple down the counter. This action is why the circuitry is called a "ripple counter." This action means that there is a period of time during which the counter could assume a variety of states as the clock is passing from stage to stage.

The count is only valid after the clock has had time to propagate through each stage. It is possible for any logic connected to the output to

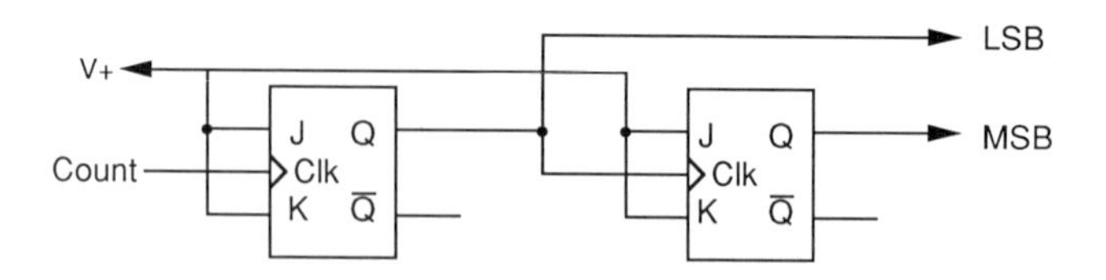

2 Bit Asynchronous Counter

(a)

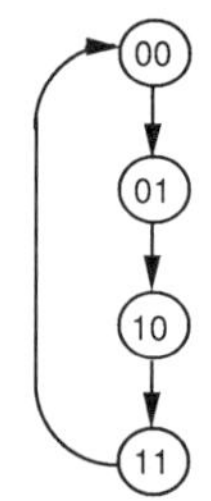

State Diagram for the 2 Bit Counter

(b)

Figure 2-15
Two Bit Asynchronous Counter.

detect a false state during this settling action. When the clock inputs of the flip-flops are not all driven by a common source, the design is said to be *asynchronous*.

The problem with the clock delay for the circuit in Figure 2-15 can be avoided by clocking all of the flip-flops *at the same time*. This would form a *synchronous counter*. That is, each flip-flop would change state in synchronization. This concept will be developed further in the section on state machines.

We now take a look at a simple circuit design. The design requirement is that we need a simple push button–activated counter. The counter is to count from 0 to F_H. The output is displayed as one digit on an LED seven-segment readout. For this section, we implement the circuit with conventional 74HC series parts. In later chapters we will look at how this requirement can be met with various programmable logic designs.

Figure 2-16 is the schematic for our circuit. The activating switch is a SPDT type. Normal mechanical switches exhibit considerable "bounc-

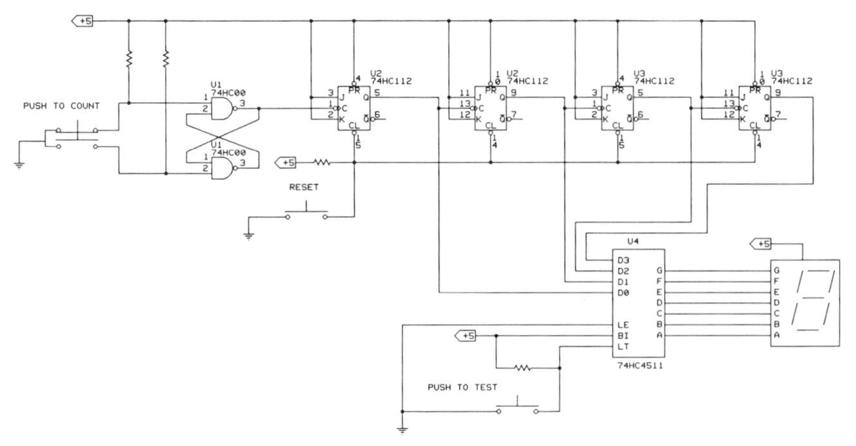

Figure 2-16
Push Button Counter.

ing'' when activated. This bouncing makes a single push of the switch appear as many closely spaced closings. Since this would cause a false reading of how many times the button was pushed, we must *debounce* the switch. In our circuit, debouncing is done with the use of a trusty RS latch.

The debouncing latch is formed by cross-coupling two NAND gates. Note the pull-up resistors on the input of the NAND gates. These are required particularly with CMOS parts to keep the input to the latch stable. Without the pull-up resistors the inputs would float low, and the latch would not function properly. To understand how the debounce circuit works, assume the switch is in the normally closed state as shown in Figure 2-16. U1A will have its input at ground so its output will be a 1. This 1 value goes to one input of U1B. The other input of U1B is held high by the pull-up resistor. U1B's output must then be 0. This 0 reinforces the 0 on the input to U1A.

When the button is first pushed, the ground connection with the upper NAND gate will be broken, and the input to the gate will go high. The output of U1A will remain high however, since the other input to the gate is still low. This state will stay in effect during the time the switch is moving to the lower contact. When the switch first makes contact with the lower contact, the situation is reversed and the state of the latch will reverse. No matter how many times the switch now bounces against the lower contact, the latch will stay in the same state until the button is released and the switch returns to the normally closed position.

The output of the latch follows the state of the switch, ignoring any multiple bounces that the mechanical action may cause or any arcing that may occur from the contacts. This stable output is used as the input to the three J-K flip-flops that comprise the counter. The output of the counter is fed to a chip that decodes the binary state of the counter into the appropriate drive signals for the seven-segment display. A switch is attached to the reset inputs of each of the J-K flip-flops. This allows us to force the counter to zero.

The circuit is simple, but it forms a useful baseline for comparing a variety of techniques for implementing both combinatorial and sequential logic functions.

We should point out some limitations to this design. The application of a ripple counter is acceptable in this application since any delay in the state changes will not be noticeable on the display. As a matter of good design practice, one should be wary of asynchronous designs. Of more importance is the fact that while our counter will count from 0 to F_H, it will only display from 0 to 9. These limitations will be overcome in the programmable implementations we will be seeing later.

2.3 Chapter Summary

To sum things up, we have looked at the three basic techniques of programmable logic: tabulation, emulation, and computation. We have covered the types of gates, logic equations, and how a "sum of products" form of an equation can be transformed into a "product of sums" form by the use of De Morgan's theorem. The concept of states has been introduced and state circuits such as latches and flip-flops have been demonstrated.

This is a lot of material, and our discussion makes no pretense of being exhaustive. The purpose was to refresh old memories as needed and introduce standard nomenclature that will be used in the rest of this book. If this section has not clicked, there are many good books on basic logic design which cover this material in greater depth. The suggested reading list in Appendix C contains several sources.

- Any logic function can be expressed in either a sum of products or a product of sums form.
- Sum of product and product of sum forms can be interchanged by the use of De Morgan's theorem.
- The maximum number of product terms (minterms) required to implement an arbitrary function of n variables is 2^n. This may require at most $n - 1$ summations.
- The maximum number of summation terms (maxterms) required to

implement an arbitrary function of n variables is 2^n. This may require at most $n - 1$ products.

- Any binary logic function can be implemented with a two level network.
- Any binary logic function can be implemented by simply ORing selected minterms together.
- When it is not clear whether a signal will need inverting, an XOR gate may be used as a "programmable inverter." This can greatly aid in efficiently implementing logic functions.

3

Combinatorial PLDs

In this chapter we will be discussing combinatorial programmable logic. We will look at each of the three major architectures for combinatorial design: the programmable read only memory (PROM), programmable array logic (PAL), and programmable logic array (PLA). Our discussion will look at how these devices fit into the standard theories of logic design, the basic architectures of the devices, and some of the more popular devices represented by each of the architectures.

The conceptually simplest form of programmable logic is the PROM, and we will begin with it. The PAL is discussed after the PROM. Following the PAL we will discuss the more complex but more flexible PLA. And finally, we will present a comparison of the three architectures.

If we look at a time line of programmable logic, the first truly programmable logic devices were PROMs that were used as design expedients. As we will see however, they do have inherent limitations in this capacity.

In order to overcome the PROM's limitations, Signetics introduced the *field programmable logic array* (FPLA) in 1975. The FPLA will be discussed in Section 3.3. The FPLA was not initially a large success, though it does have certain advantages in comparison to other programmable devices. In order to produce a more efficient design, MMI simplified the relatively complex FPLA to what they called *programmable array logic* (PAL). The simpler architecture and improved supporting software available with the PAL allowed it to rapidly gain dominance over the PLA.

Before we talk about actual PROMs, PALs, and PLAs, we will take a few moments to briefly discuss the basic mechanism by which information is stored in PLDs. The two basic mechanisms used as storage elements in PLDs are the *fuse* and the *floating gate* cell. Various vendors make use of different technologies and techniques to fabricate fuses and floating gates. The following discussion is based on TI's processes for fabrication of PLDs.

In general, bipolar technologies such as TTL make use of fuses as the basic storage element in programmable circuits. MOS technologies such as CMOS make use of floating gates.

The fuse used in programmable logic circuits is essentially the same as the standard equipment fuse that one is familiar with from everyday use. The major difference is that the fuse is fabricated on a microscopic scale, thus allowing thousands of them to be placed on a single IC.

Figure 3-1a shows the layout of a TiW (titanium–tungsten) bipolar fuse in the unprogrammed state. The conductive metal is composed of TiW approximately 500 Å thick. TiW is also used as a barrier material to prevent the aluminum in the IC from coming into contact with the silicon substrate.

The fuse is programmed by passing a high current through the fuse. This high current causes the fuse to heat up to approximately 2100°C. At this temperature the TiW metal becomes molten. At this point, two things happen: First, the molten metal flows thus breaking the contact. Second, the silicon dioxide (SiO_2) above the fuse melts and flows into the gap. The result is a very positive insolation. The programmed fuse is shown in Figure 3-1b.

Over the years CMOS technology has been gaining ground as the technology of choice. For use in CMOS circuits, the metal fuse is replaced with a floating gate avalanche injection MOS (FAMOS) transistor.

The FAMOS transistor resembles an ordinary MOS transistor. The major difference is the addition of a floating gate buried in the insulator between the substrate and the ordinary select gate. This structure is shown in Figure 3-2a. The schematic symbol for a FAMOS transistor is shown in Figure 3-2b.

The FAMOS transistor is programmed by capacitively coupling the select gate in series with the floating gate. By a process called *hot electron injection,* a charge is placed onto the floating gate. This electron injection is caused by pulling the select gate to the programming voltage and the drain of the FAMOS transistor to the programming voltage minus several threshold drops. Once a charge has accumulated on the gate, the threshold voltage of the transistor is shifted. This shifted threshold of the transistor provides a switch action similar to that of a fuse.

Once programmed, the FAMOS transistor retains the electron charge until exposed to an ultraviolet (UV) light with a wavelength of 2,537 Å. The UV light will erase the charge by giving the electrons enough energy to scatter from the floating gate.

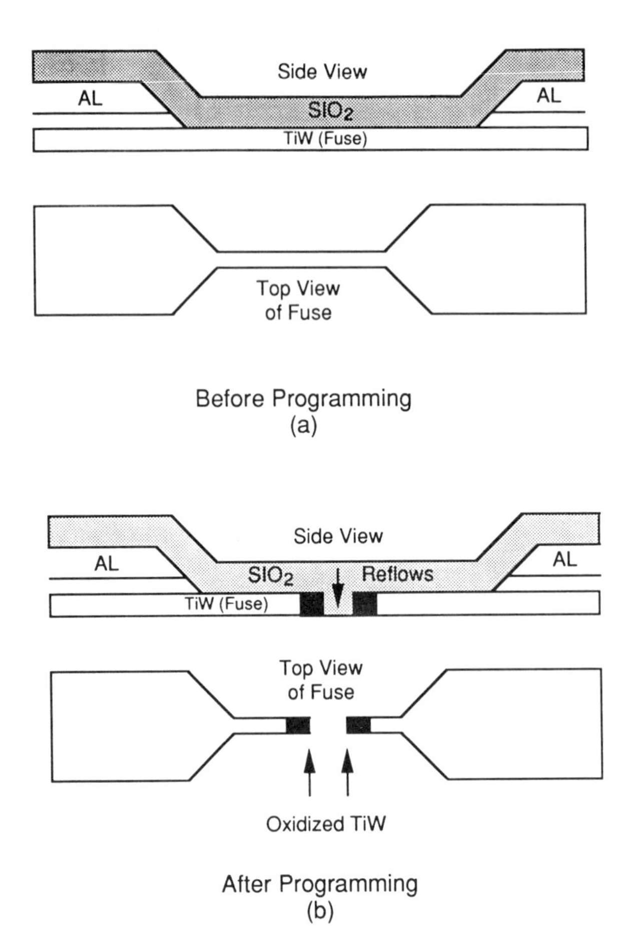

Figure 3-1
Titanium Tungsten Fuse. (Reprinted with permission of Texas Instruments, ©1990.)

This ability of UV light to return the cell to its unprogrammed state is a major advantage of CMOS designs for two reasons. First, the erasable nature of these devices allows 100% testing of the device during manufacture. Obviously this is impractical with fuse-based designs. Second, if the IC is provided in a windowed package, the PLD can be erased and reprogrammed by the user.

An interesting variation on the floating gate cell is the electrically erasable floating gate. Operation is similar to the floating gate but, as the name implies, the cell can be erased by applying an erase voltage instead of UV light.

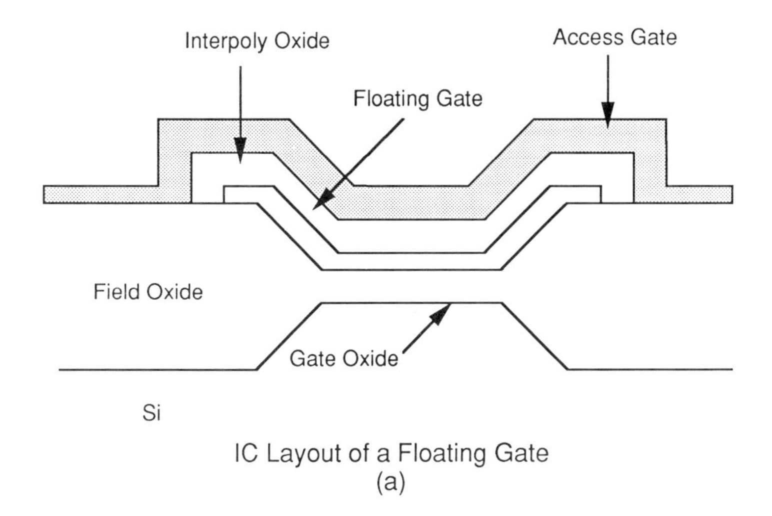

IC Layout of a Floating Gate
(a)

Schematic Symbol for a Floating Gate
(b)

Figure 3-2
FAMOS Floating Gate Cell. (Reprinted with permission of Texas Instruments, ©1990.)

3.1 Programmable Logic and the PROM

When used as programmable logic elements, PROMs are sometimes called programmable logic elements (PLEs). Calling PROMs "PLEs" confuses the issue some, but it does differentiate their use as programmable logic from their more conventional role as memory elements. For our purposes, the terms PLE and PROM are interchangeable, and we will defer from any semantic arguments over which term should be used when.

In Chapter 2 we defined the three basic ways programmable logic functions are implemented: *tabulation, emulation,* and *computation.* PROMs inherently fill the bill for the tabulation option.

In a conventional computer application, PROMs are used to permanently store program code and data that is unchanging. Typical examples include boot routines for computers and character translation tables. In these applications the use of the PROM is straightforward:

1. An address is applied to the PROM to select the data word that is to be read.
2. The PROM is enabled, thus allowing the data word to be placed on the data bus.
3. The PROM is deselected and no longer drives the data bus.

The architecture of a typical PROM is shown in Figure 3-3. The high-order five bits of the address are applied to a *row decoder*. As the name implies, the row decoder selects one of the 32 rows in the programmable array. The lower four bits of the address are applied to the eight *1 of 16 multiplexers*. The combination of high-order and low-order bits select one eight bit word from the array. The output of the multiplexers are buffered by relatively high-current, three-state devices. The two chip controls, /RD and /CS, enable the output of the buffers. When used as a PLD, the /CS and /RD lines of the PROM are typically tied low so that the output will always be enabled.

When PROMs are used as a PLD, the PROM is generally wired to be permanently enabled. The address pins become the *logic inputs*, and the

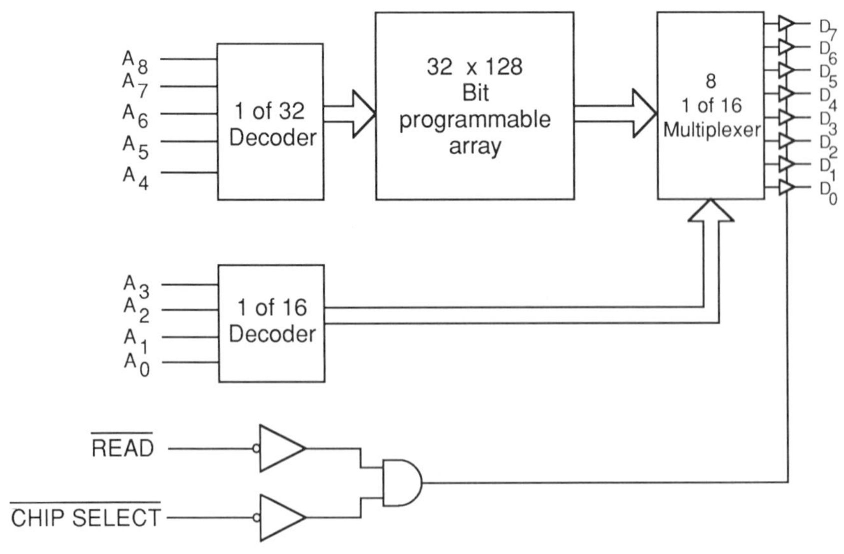

Figure 3-3
General Architecture of a PROM.

data output pins become the *logic outputs*. The data words programmed into the PROM form the *transfer function* we are trying to realize.

This is all better illustrated with an example. In Chapter 2, we presented a simple counter realized with discrete logic. One of the failings of this circuit is that it only displays decimal digits 0–9, even though the counter itself actually counts $0–15_D$, or $0–F_H$. The reason for this is that our display driver, the 74HC4511, only displays the digits 0–9. For Hex values from *A* to *F*, the display will simply go blank. There are display drivers available that will display a Hex count on a seven segment display but they are fairly rare. So for our exercise, we will utilize a PROM to realize a hexadecimal value display driver.

Figure 3-4 shows the layout of a seven-segment display and the patterns that are displayed for each of the 16 Hex values. The first step in realizing our Hex display driver is to set up a mapping from the output of the counter, through the display driver, then to the display. The mapping from the counter to the display drive is straightforward. We will simply assign D0 from the counter to A0 on the PROM (D1 maps A1, etc.). For the outputs, we will assign D0 of the PROM to segment A, D1 to segment B, etc. A schematic of our circuit is shown in Figure 3-5. The 74HC244

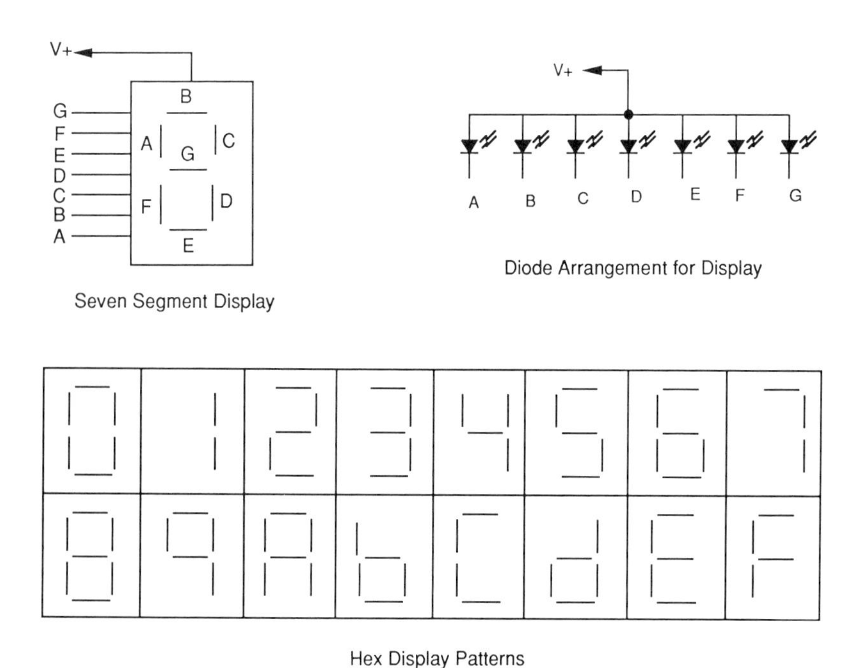

Figure 3-4
Seven-Segment Display Configuration.

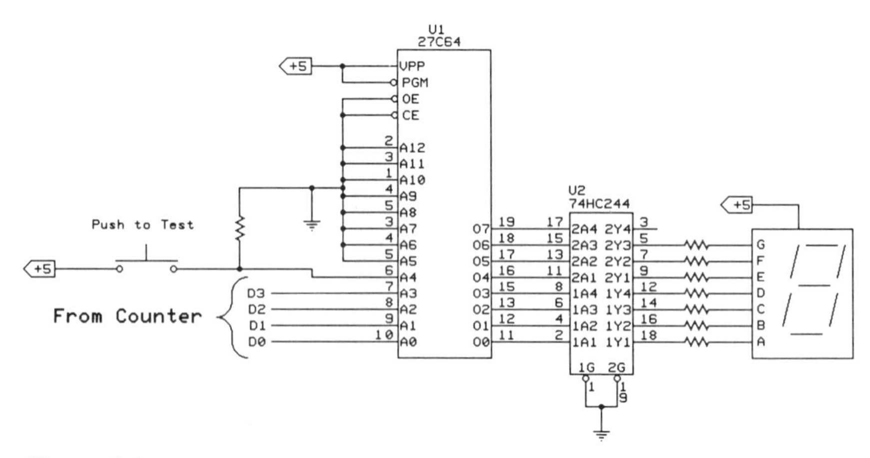

Figure 3-5
PROM-Based Decoder Circuit.

shown in this figure is there to buffer the relatively low-level drive from the PROM up to current levels that will make the LED glow brightly.

Notice that a special feature has been added to the display driver. Switch S2 has been added as a "TEST" control. The idea is to allow us to check the integrity of the display. When TEST is pushed, all of the segments will light thus giving a positive indication that all elements of the display are functional.

With the information shown in Figures 3-4 and 3-5, there is enough information to build a data table for the PROM. Such a table is shown in Figure 3-6. Notice that for our mapping of the D output of the counter to the address inputs of the PROM allows us to develop the programming by simple inspection. For example, in case 0, the input to the PROM is simply 0. For this case, we want segments A, B, C, D, E, and F turned on. To accomplish this we simply place a 0 in the corresponding columns of row 0. Segment G is turned off, so we place a 1 in the corresponding location. Our table is filled in for each input case in exactly the same way. In effect, the logic input conditions form the address input to the PROM. The logic function outputs are simply the data values stored at that address.

Later chapters will discuss the tools available for programming PLDs, but for now we will point out some options for generating the code for the PROM. The conceptually simplest is to use a conventional assembler. Almost any one would do since only the "define data" pseudo-ops

HEX Value of Input	Address					Data							
	A4	A3	A2	A1	A0	D7	D6	D5	D4	D3	D2	D1	D0
00	0	0	0	0	0	1	1	0	0	0	0	0	0
01	0	0	0	0	1	1	1	1	1	0	0	1	1
02	0	0	0	1	0	1	0	0	0	1	0	0	1
03	0	0	0	1	1	1	0	1	0	0	0	0	1
04	0	0	1	0	0	1	0	1	1	0	0	1	0
05	0	0	1	0	1	1	0	1	0	0	1	0	0
06	0	0	1	1	0	1	0	0	0	0	1	0	0
07	0	0	1	1	1	1	1	1	1	0	0	0	1
08	0	1	0	0	0	1	0	0	0	0	0	0	0
09	0	1	0	0	1	1	0	1	1	0	0	0	0
0A	0	1	0	1	0	1	0	0	1	0	0	0	0
0B	0	1	0	1	1	1	0	0	0	0	1	1	0
0C	0	1	1	0	0	1	1	0	0	1	1	0	0
0D	0	1	1	0	1	1	0	0	0	0	0	1	1
0E	0	1	1	1	0	1	0	0	0	1	1	0	0
0F	0	1	1	1	1	1	0	0	1	1	1	0	0
10	1	0	0	0	0	1	0	0	0	0	0	0	0
11	1	0	0	0	1	1	0	0	0	0	0	0	0
12	1	0	0	1	0	1	0	0	0	0	0	0	0
13	1	0	0	1	1	1	0	0	0	0	0	0	0
14	1	0	1	0	0	1	0	0	0	0	0	0	0
15	1	0	1	0	1	1	0	0	0	0	0	0	0
16	1	0	1	1	0	1	0	0	0	0	0	0	0
17	1	0	1	1	1	1	0	0	0	0	0	0	0
18	1	1	0	0	0	1	0	0	0	0	0	0	0
19	1	1	0	0	1	1	0	0	0	0	0	0	0
1A	1	1	0	1	0	1	0	0	0	0	0	0	0
1B	1	1	0	1	1	1	0	0	0	0	0	0	0
1C	1	1	1	0	0	1	0	0	0	0	0	0	0
1D	1	1	1	0	1	1	0	0	0	0	0	0	0
1E	1	1	1	1	0	1	0	0	0	0	0	0	0
1F	1	1	1	1	1	1	0	0	0	0	0	0	0

Figure 3-6
Data Table for a Seven-Segment Decoder.

are needed. For example, if "DB" stands for "define byte," a partial listing of the code might look like this:

```
              ORG 00          ; Start at address 00
0000 81  DB   11000000   ; Pattern for Count 0 input.
0001 E7  DB   11110011   ; Pattern for Count 1 input.
0002 C8  DB   10001001   ; Pattern for Count 2 input.
                   *
                   *
                   *
000F 9C  DC   10011100   ; Pattern for Count F input.
```

The object file generated by the assembler is used to program the PROM, exactly as if we were generating a conventional program to be burned into a PROM.

A second option would be a special type of assembler called a meta-assembler. A meta-assembler's listing would look similar to the assembler example we just presented. The meta-assembler has the added advantage of being able to define bit fields and our own pseudo-ops. The output of a meta-assembler is called *microcode*. A microcoded listing might look like this:

```
          ORG 00                                          ; Start at address 00
0000 81   SEGMENT X, OFF, ON, ON, ON, ON, ON, ON          ; 0 input.
0001 E7   SEGMENT X, OFF, OFF, OFF, ON, ON, OFF, OFF      ; 1 input.
0002 C8   SEGMENT X, ON, ON, ON, OFF, ON, ON, OFF         ; 2 input.
                *
                *
                *
000F 9C   SEGMENT X, OFF, OFF, ON, ON, ON, OFF, OFF       ; F input.
```

The final option available is to input a table, very similar in form and content to Figure 3-6, into a PLD compiler. Most compilers can take a table like this and convert it to a data pattern for a PROM or other PLD. We will look at how to use PLD compilers in Chapter 5.

Two important points should be made with respect to this example. Referring to Figure 3-6 again, notice that it was not necessary to define any logic equations. We did not need and are not interested in the transfer function. This is an advantage of tabulation techniques: we only need to know what outputs we want for each input. There is no need to generate, check, and simplify logic equations to achieve our desired end.

The second important point can be illustrated by looking at the TEST input. The display might be in any state when we perform the display test. To accommodate this condition we must use not only the original 16 bytes to encode the display pattern, but an additional 16 bytes for the test function. This is true even though all of the second group of 16 bytes contain the same value of 80. This is true in general for PROMs; for each additional input the size of the PROM must double. Or in other words, the size of the PROM must be 2^n, where n is the number of inputs. This doubling of size for each extra input comes about as a function of the architecture of a PROM. We will revisit this when we compare PROM, PAL, and PLA architectures.

A more common use of PROMs as PLDs occurs in address decoding applications. This application of PROMs spurred much of the interest in programmable logic and therefore is worth looking at closer. To understand the motivation behind using a PROM as programmable logic, consider the design environment in the mid-1970s. This was a very exciting period for the electronics industry. Circuit design techniques were rapidly changing. New and sophisticated semiconductor components were being announced almost daily.

This rate of change was not entirely a blessing, however. Consider the fate of the designers of memory boards. The capacity of memory chips was doubling at a rate of once every six months to a year. Normal product introduction time for a memory board was also six months to a year. This meant that any new product was likely to become obsolete just about the time it *entered* the market! Naturally, this gave nightmares to product planners. A new definition of obsolete was generated: "If we could build it, it was obsolete!"

Product planners adopted a variety of strategies to deal with the problem. Some chose to simply design their boards around chips that were announced but not yet in production. This was risky since product introduction had to be delayed until the memory chips became available. If the chips were not ready by the time the boards were ready for production, a large investment in time, money, and materials was simply left idling. Others decided to build with existing chips and risk putting board products on the market that were obsolete as soon as they were introduced.

Neither of these techniques were acceptable. A better way had to be found. The semiconductor manufacturers provided part of the solution by designing their memory parts to follow a standard form factor. It was agreed that future parts that were not yet buildable would follow a standard pin convention. The specifications were standardized under JEDEC conventions. Board producers could now at least anticipate what the pins would be when their 256×8 bit IC was upgraded to a 512×8 bit IC. Boards could be designed to use the existing parts, but be upgradable to higher densities by simply changing out the parts on the board.

As we noted however, standardizing pin outlines was only part of the solution. What was needed was a simple and efficient way of installing a decoding network that would meet any of the possible memory configurations. Remember, the board was designed to be upgraded by removing lower density chips and installing higher density chips. The ideal solution would be to simply make the decoding network change part of the chip change-out procedure. Thus a new decoding chip could be plugged in to meet any new memory requirement. But how can one come up with such a chip in a simple way? The answer was as simple as it was ingenious: "Don't use a logic chip, use a memory chip!" Specifically, use a PROM.

This elegant solution to the decoding problem gave rise to the term "decoder PROM." For several years, virtually every memory board produced used some variation on this technique. Modern designers are more likely to use one with more sophisticated programmable logic solutions (which is what the rest of the book is about), but the term "decoder PROM" remains in the lexicon of the digital designer to this day. Later in this chapter we will look at a typical decoder application.

The success of the PROM in decoding led to consideration of it in other logic roles, such as our display driver example. When viewed as a logic element, rather than as memory element, the PROM can be viewed as a vehicle for implementing a classic minterm expansion of an arbitrary logic function in n variables, where n is the number of address lines into the PROM.

Consider, for example, $y = f(a, b)$. There are two input variables (a and b), so f can be at most $2^2 = 4$ terms in complexity: 00, 01, 10, and 11. If we imagine a 2×1 PROM, we can arbitrarily program the output to be whatever we want for each and every one of the four possible input conditions. We can expand this example to any realizable PROM. This can be seen by considering the minterm/maxterm expansions discussed in Chapter 2. Further, while it may not be intuitively obvious, this argument provides formal proof that any logic function of an arbitrary number of levels can be reduced to an implementation with only *two* levels. This will probably make more sense when the three basic PLD architectures are compared.

For all this flexibility, the PROM never caught on as a replacement for combinatorial logic for several reasons. First, the number of inputs (i.e., address lines) are fixed. Similarly, the number of output lines (i.e., data lines) are also fixed. This makes it difficult to fit the PROM into a wide range of applications. Pins are a valuable resource on a chip, and if they cannot be used efficiently, design costs increase. Second, as noted previously, the size of the PROM doubles with each variable added to the function. Our simple two variable example only required a four word PROM. Three variables would have required eight words. This is not too bad, but consider going from eight variables to nine variables; this requires going from 256 words to 512 words!

Further, in the real world a larger number of product terms are available in a PROM than are usually required. Studies have shown that approximately 70% of the functions implemented with PLDs can be implemented with three or fewer product terms.

What was needed was a device that would retain the flexibility of the PROM but implement logic functions more efficiently. The solution was, at least in part, the PLA. Since the PLA is more complicated than the PAL, we will first look at the PAL.

3.2 Programmable Array Logic

The second option for realizing programmable logic functions, as discussed in Chapter 2, is emulation. In emulating a discrete logic design, we achieve the equivalent of a function by selecting product terms in a standard sum of products (SOP) form. The PLD emulates, functionally and logically, what we would get if we were to wire individual gates together. As discussed in Chapter 2, any combination of gates can be reduced to a simple two layer network. The individual terms may become "wider," but no matter how "deep" the network it can be reduced to only two levels. These comments apply to the PROM, the PAL discussed in this section, or the PLA to be discussed in the next section.

To understand the fundamental ideas behind the PAL, consider Eq. (2-3) from Chapter 2. For convenience it is repeated here as Eq. (3-1):

$$F = \overline{A} * \overline{B} * \overline{C} + A * \overline{B} * \overline{C} + \overline{A} * B * \overline{C} \tag{3-1}$$

As we discussed in Chapter 2, this is not the simplest form of the function. It is in a canonical form, however.

Figure 3-7 is a schematic of one way to realize a circuit that implements the function expressed by Eq. (3-1). Note that the realization is a

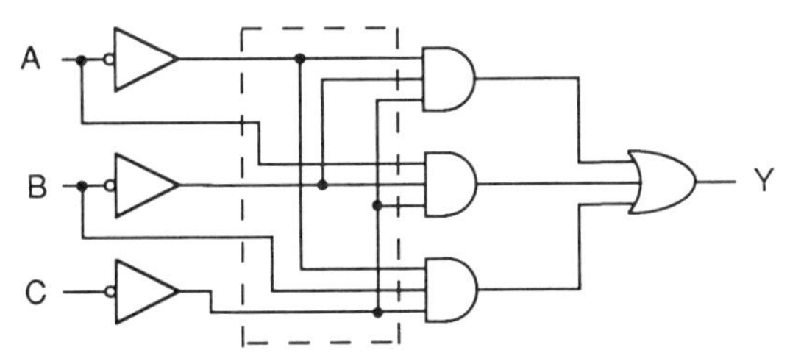

Figure 3-7
One Implementation of Eq. (3-1).

direct one-for-one mapping between the function and the circuit. No attempt at the simplification that we know is possible has been made.

Figure 3-8 is logically the same as Figure 3-7 but it has been redrawn to emphasize several key points. Let us examine the area outside the dashed box. Notice that the area to the right of the dashed box is independent of the specific values of the expressions in Eq. (3-1). This portion of the circuit can be viewed as simply implementing the minterm expression $m_x + m_y + m_z$, where x, y, and z could be any minterm.

Note that the input circuitry (now drawn at the top of the dashed box) is also independent of any particular function. The input circuitry simply provides both the true and complement values of the inputs.

The only portion of our circuit that is unique to the particular function is the interconnect section located inside the dashed box [Eq. (3-1)]. The point is that the circuit as it is drawn can compute *any* function of three variables having no more than three product terms. The *only* things that determine which function is computed are the interconnects. This is the key concept behind the PAL. By providing a general architecture with a programmable interconnect, we achieve a very flexible programmable logic circuit.

As noted earlier, analysis has indicated that approximately 70% of the applications for a PLD can be implemented with three or fewer product terms. So far so good, but what if our function requires more than three terms to implement? There are two answers. First, as we will see shortly, a wide variety of PALs are available. These have different combinations and sizes of AND gates and OR gates to accommodate a variety of functions. Most PAL architectures provide for approximately eight sums per product term.

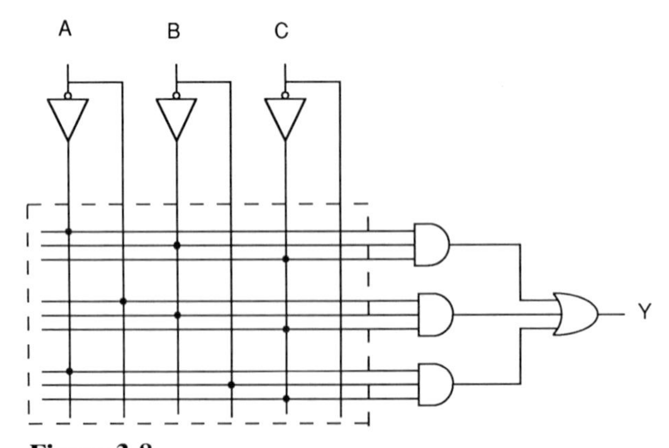

Figure 3-8
Figure 3-7 Redrawn to Emphasize Interconnects Array.

If a PAL cannot be found that will meet all of the requirements of the functions, a second possible answer is to use a PLA, which is covered in Chapter 4.

Figure 3-8 provided a good conceptual model of a PAL architecture. For most applications it is not necessary to dive any deeper into the actual layout of the PAL. One simply specifies the equations, and the actual mapping is handled by the software. This is a generally unsatisfying explanation for the engineer who is truly interested in how a PAL is structured internally. Further, it is occasionally necessary to get involved with the actual fuse maps of a PAL. For these reasons we will take a closer look at the internal circuitry of the PAL. From this detailed view we will demonstrate how simplifying conventions for drawing PLDs have been developed.

Figure 3-9 shows a general purpose implementation of a very simple PAL. As noted earlier, PLDs were based on the technologies available for handling PROMs. At the time of the first PALs and PLAs, this meant that

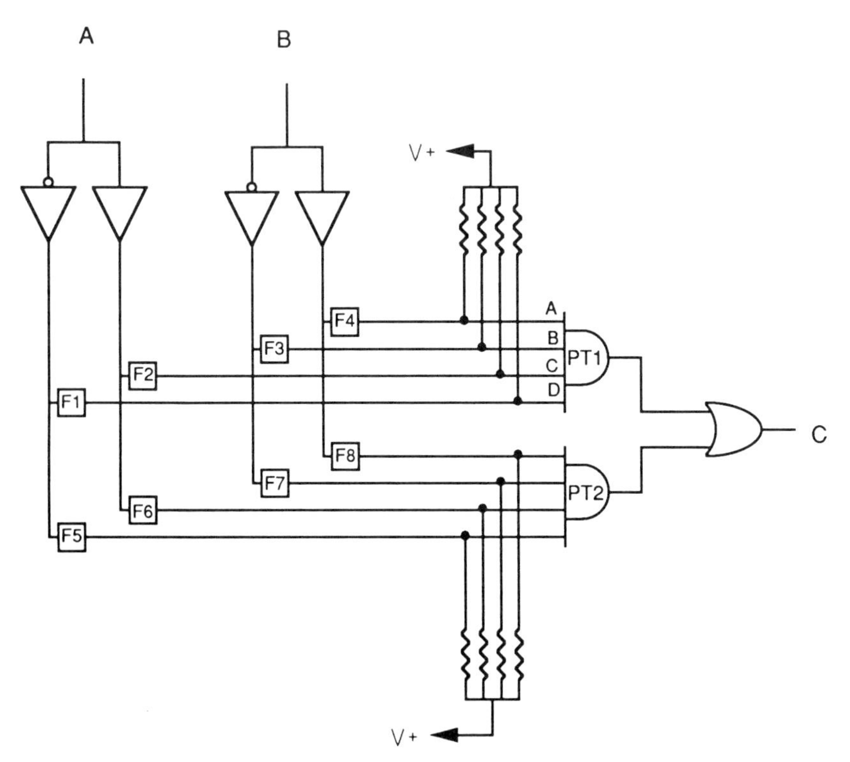

Figure 3-9
Detailed Schematic of a Simple PAL.

the only viable technique for a programmable device was the microfuse. Since fuses are inherently one-time-breakable devices, the fuse itself dictated the detailed layout of a PAL circuit.

Notice the two product terms PT1 and PT2 in Figure 3-9. These are AND gates with pull-up resistors on the inputs. Each input to the product term is connected to either the true or complement value of each of the inputs. Thus the size of the AND gate must be twice the number of inputs. The two product terms are summed by an OR gate.

When the PAL is manufactured all fuses are in place. The programming process consists of passing high current levels through those fuses that are to be opened.

In order to get a feel for how this whole thing works, let us look at the fuse-pair F1 and F2. These fuses are associated with product term 1 (PT1) and the signal *A* input. Since there are two fuses, each of which can be either open or closed, there are four possible cases:

Case 1: Both fuses are intact This is the case at the time of manufacture. In this case, both the true and the complement of *A* will be applied to PT1. Since *A* AND /*A* is 0 regardless of the value of *A*, PT1 will contribute nothing to *C* (i.e., PT1 is a "don't care" as far as the output *C* is concerned). Notice that this is true regardless of what is done with the fuses for the *B* input.

Case 2: F1 open, F2 intact In this case, the inverted portion of signal *A* is disconnected from PT1. The input PT1D is pulled high by the pull-up resistor, thus making the /*A* a "don't care" for PT1. With F2 left intact, the true sense of signal *A* is applied to PT1.

Case 3: F1 intact, F2 open This is, of course, just the opposite of Case 2. The complement of signal *A*, /*A*, is applied to PT1.

Case 4: Both fuses are open With both fuses open, signal *A* is completely disconnected from PT1. PT1's output will only be a function of *B*.

The same sort of analysis can be applied to each of the other fuse-pairs.

To see how we would actually program the simple PAL, let us look at some specific functions. To start with, let us take the case of building a simple INVERTER, *C* = /*A*. To simplify the presentation, the fuses will be listed on one line. To indicate an intact fuse, an "X" is used, and to indicate an open fuse a "–" is used.

```
Fuse  1 2 3 4 5 6 7 8
      X - - - X X X X
```

This connects the inverted signal *A* to PT1. All other inputs to PT1 are pulled high by the pull-up resistors. Thus PT1's output will simply be the inverted signal *A*. PT2's output will be 0 since all of its fuses are intact. Thus, the output *C* will simply be an inversion of whatever is on *A*.

Another interesting case is $C = A$ AND B. This fuse map would be:

Fuse	1	2	3	4	5	6	7	8
	–	X	–	X	–	–	–	–

In this case, PT2 does not contribute to C.

A more interesting case is $C = A$ XOR B. We cannot directly implement an XOR operation but if we remember that

$$A \text{ XOR } B = (A * \overline{B}) + (\overline{A} * B)$$

we can accomplish our goal.

The map for the XOR function is:

Fuse	1	2	3	4	5	6	7	8
	–	X	X	–	X	–	–	X

In this case both PT1 and PT2 are contributing to the output C.

Originally PLDs were programmed in exactly this manner. The engineer was required to generate a fuse map with a simple editor. This process is tedious and error prone, naturally. Modern PLD software frees the engineer of this responsibility. All that is necessary is to write the equations. For example, the line

$$C = (A \ \& \ /B) + (/A \ \& \ B)$$

in the PLD source file will generate the XOR of the inputs A and B on the output C. The fuse map is automatically generated by the software.

The schematic in Figure 3-9 is clear and easy to understand, but for large numbers of gates, inputs, and outputs it would get very cluttered. For this reason, several simplifying conventions have been adopted for drawing PLDs.

The first simplifying convention is to drop the pull-up resistors. Any line that is not driven by a fuse can be assumed to go to its innocuous (not inactive!) state. For AND gates this means a logical 1. For OR gates this means a logical 0.

The second simplifying convention is to show the true and complement input buffers as a single differential driver.

A third convention, which would seem rather strange at first glance, is to eliminate the fuses from the drawing. This is not as strange as it sounds, since it is clear in the drawings where the fuses are (generally at the intersection of lines). In drawing PLDs, if we want to show which fuses are left intact for a particular application, we simply show an "X" at the point where a fuse would have been drawn. Where it is not clear that a fuse is implied, the fuse is explicitly drawn. This is often the case when special features of a PLD are programmable by a special fuse.

The final convention is to not draw all of the inputs for each of the product terms. The individual lines are replaced with one long line. These conventions are adopted in Figure 3-10, which is simply Figure 3-9 redrawn with these conventions.

PLD diagrams can be very confusing to those who have not had experience interpreting them. If the reader gets confused when looking at PLD architecture drawings, it helps to remember that these simple conventions have been applied. The sometimes seemingly nonsensical ways the drawings have been produced is a simplified shorthand that makes it easier to show the key and unique features of a particular architecture.

Before looking at some typical examples of real PAL architectures, it is worth additional comment on one of the key features of the PAL, the fixed OR output. Notice that we have control of how the product terms are used: We can drive them from a variety of inputs, set all the inputs to a logical 1 (by opening all the input fuses), and essentially make them disappear by feeding them a contradictory input (*A* AND /*A*, the unprogrammed state). But for all this control of the product terms, we have no control of the OR, or summation, terms. The summation terms are inherently fixed as a function of the particular PAL architecture. Whether this is a disadvantage or not depends upon several factors, including the particular application. To fully understand the trade-offs, we need to look at the PLA architecture in the next section. After doing this, we will compare and contrast PALs versus PLAs. But for now let us look at several of the classic PAL architectures.

Figure 3-11 shows one of the most popular PALs, the 16L8. The one shown is a TI part but this device is manufactured by a variety of vendors. Notice that there are two dedicated outputs and six pins that can be used as either inputs or outputs.

The PAL16L8 is often used in address decoding applications due to its flexible architecture and active low outputs. These features allow the

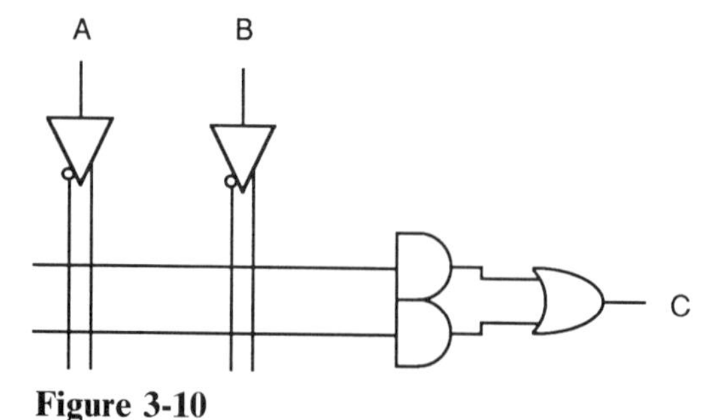

Figure 3-10
Figure 3-9 Redrawn with PLD Conventions.

PAL16L8AM, PAL16L8A-2M
STANDARD HIGH-SPEED *PAL*® CIRCUITS

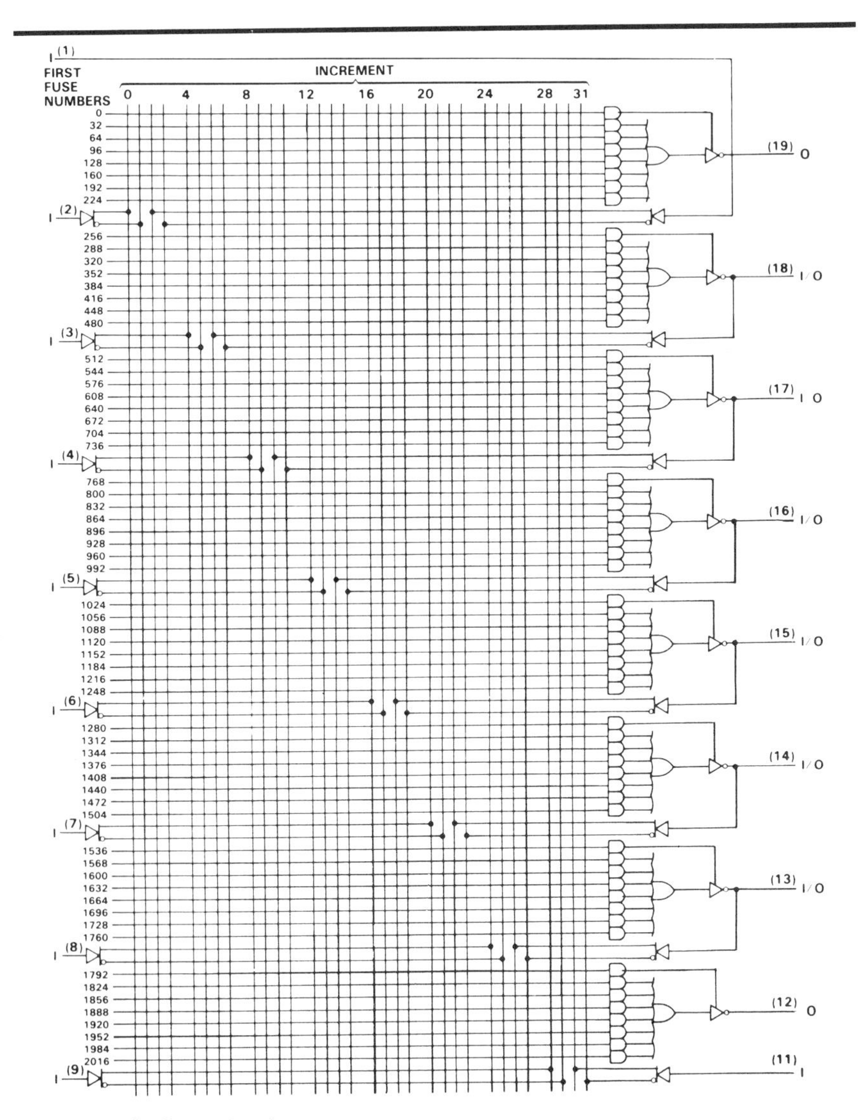

Fuse number = First Fuse number + Increment

Figure 3-11
The PAL16L8. (Reprinted with permission of Texas Instruments, ©1990.)

PAL16L8 to do the address decoding for a variety of microprocessors. The PAL16L8 is also often a good fit for combining many "glue" gates into a single device.

Figure 3-12 shows a variety of common combinatorial PALs. Notice that there is a bias toward active low or inverted output architectures. It is not necessary to understand why this is the case to be able to use PALs, but since it is a bit of interesting engineering trivia that still impacts design practice, we will take a few minutes to explore it.

Due to the basic nature of silicon, it is easier to make a NPN transistor than a PNP transistor. Thus, the first popular IC logic family, the resistor–transistor logic (RTL) family, was composed almost entirely of NPN transistors. The NOR gate is the simplest gate in the RTL family and is shown in Figure 3-13. Biasing these devices in their common-emitter configuration required that the load resistors and the base resistors be tied

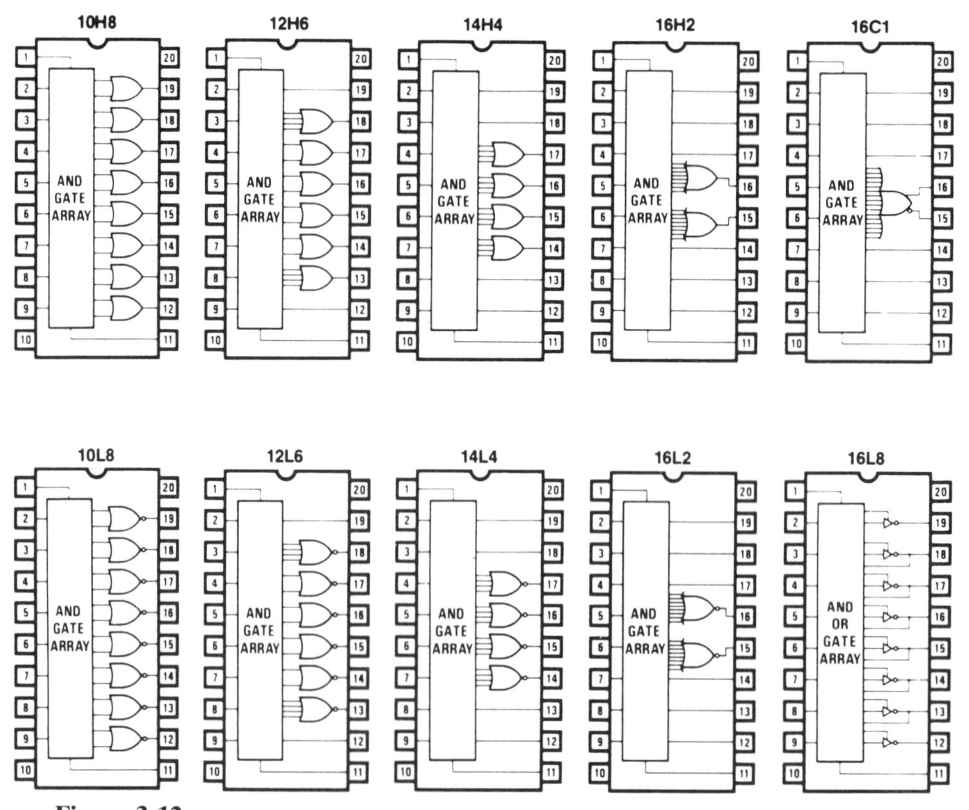

Figure 3-12
Some Common PAL Architectures.

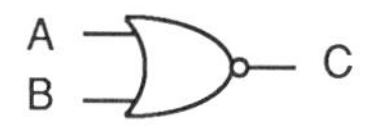

Logic Symbol for a NOR Gate
(a)

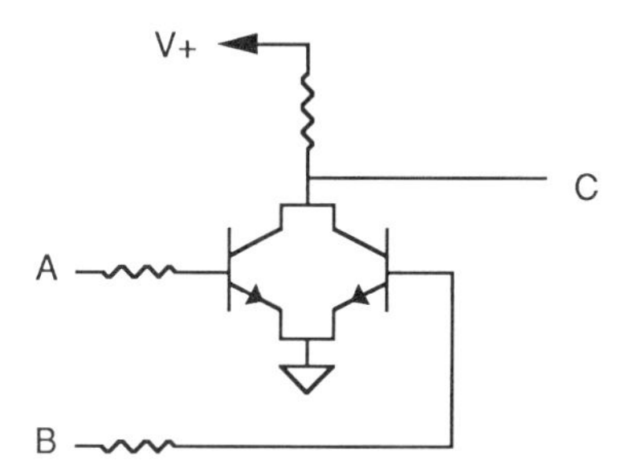

Schematic for a RTL NOR Gate
(b)

Figure 3-13
Classic RTL NOR Gate.

to a positive voltage. A natural result of this architecture was that inverting devices, such as NOR gates, were faster than noninverting devices, such as OR gates, since they did not require the addition of an inverter on the output. Furthermore, the transition time during switching was much faster for devices going to 0 (ground) than for devices switching to one (>2 V). This was the case since the transistor actively pulled the signal low, but only the load resistor was available to pull the signal high.

A resistor is often used to bias the input of a gate to a default value. Since such a resistor must be compatible with the load resistor inside the driving gate, these resistors were typically used to "pull up" the inputs. Such a "pulled up" node can be driven with multiple gate outputs, thus resulting in a "wired-OR" node.

For the above reasons, the control inputs of peripherals such as memories are typically active low signals. Since PLDs are often used as address decoders or control decoders, it is natural that they tend toward having an active low output. Theoretically, this active low output for PLDs does not make much difference. But from a practical point of view, it always leads to confusion in the programming process. We will look at this more carefully in Chapter 4.

For completeness of the discussion, we also point out that these factors have essentially disappeared for modern logic. Most new designs are based around CMOS devices that have very symmetrical properties for active high or active low transitions. More designs are appearing that make use of pull-down resistors as well as the traditional pull-up resistors. Active high and active low signals are mixed much more uniformly.

Most modern PLDs reflect this and are capable of either active high or active low outputs. We will see examples later in this chapter. For the near future, however, one can expect to see active low control signals, and the PALs shown in Figure 3-12 are far from being obsolete.

Finally, the PAL is often described as a programmable AND array/fixed OR array device. This description follows from its basic architecture.

3.3 Programmable Logic Arrays

As noted in the introduction to this chapter, PLAs were the first devices dedicated to the idea of programmable logic. PLAs are based on the classic implementation of minterm and maxterm equations.

Unlike PALs, PLAs have both programmable AND arrays and programmable OR arrays. This flexibility allows more general and efficient utilization of the on-chip resources. Consider our example of the INVERTER in the last section. Even though the INVERTER only made use of the one product term, it still required the full set of resources associated with the output *C*. These included PT1, PT2, and the OR gate. Our simple INVERTER required exactly the same resources as did the far more complex XOR function. For real world PALs such as the 16L8, the inefficiency would have been even greater. This is true in general for PALs: The amount of on-chip resources tied up are a function of the device architecture, not the complexity of the function realized. Simple functions use the same percentage of resources as do complex functions.

Now let us look at the same function realized with a PLA. Figure 3-14 is a simplified diagram of a typical PLA. Notice that the input and the product terms are the same as for the PAL. The big difference is that instead of the product terms being connected to a fixed OR gate, they are instead connected to another programmable array. This second programmable array is connected to a series of OR gates. The OR gates provide the summations terms, so we have labeled them ST1 and ST2.

We have shown in Figure 3-14 the necessary connections to implement a simple INVERTER. Notice that this implementation required only

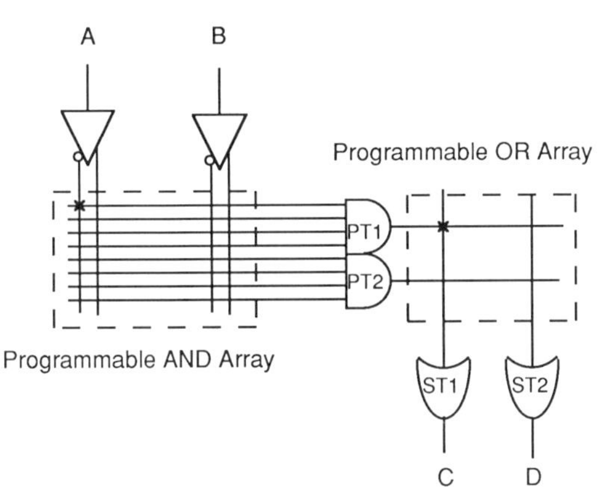

Figure 3-14
Simplified PLA.

one product term and one summation term, as opposed to the two product terms and one summation term required by the PAL implementation.

The advantages of the PLA architecture can be very great in complex designs. For example, address decoders often have product terms in common for a variety of outputs. In the PAL, this commonality cannot be exploited effectively. In the PLA, any product term that is common can be shared among outputs via the programmable OR array.

From a theoretical perspective, both the PAL and PLA architectures implement a SOP form of equation. Product of sum (POS) equations must be converted to SOP form before they can be implemented. For the PLA, this is often a far more efficient process than for the PAL. The reason for this is the ability of the PLA to share common terms.

A very popular PLA, Signetics' PLS-153, is shown in Figure 3-15. Notice that the PLS-153 has not only the conventional AND and OR programmable matrixes but also a few additional interesting features. The first of these are the programmable inverters (S_0 through S_9) on the outputs. In Chapter 2 we discussed how an XOR gate can be used as a programmable inverter. The S terms in the PLS-153 are a classic example of the implementation of programmable inverter in PLDs.

A second interesting feature of the PLS-153 is the three-state output buffers. Each of these buffers is under control of its own product term. Thus individual outputs can be driven high, driven low, or effectively disconnected from the external circuit. The final interesting feature of the PLS-153 is its feedback terms from the I/O pins. This feedback capability can be used in one of two ways. First, it can be used to implement latches and registers. Second, the I/O pins can be used as inputs. This allows, for

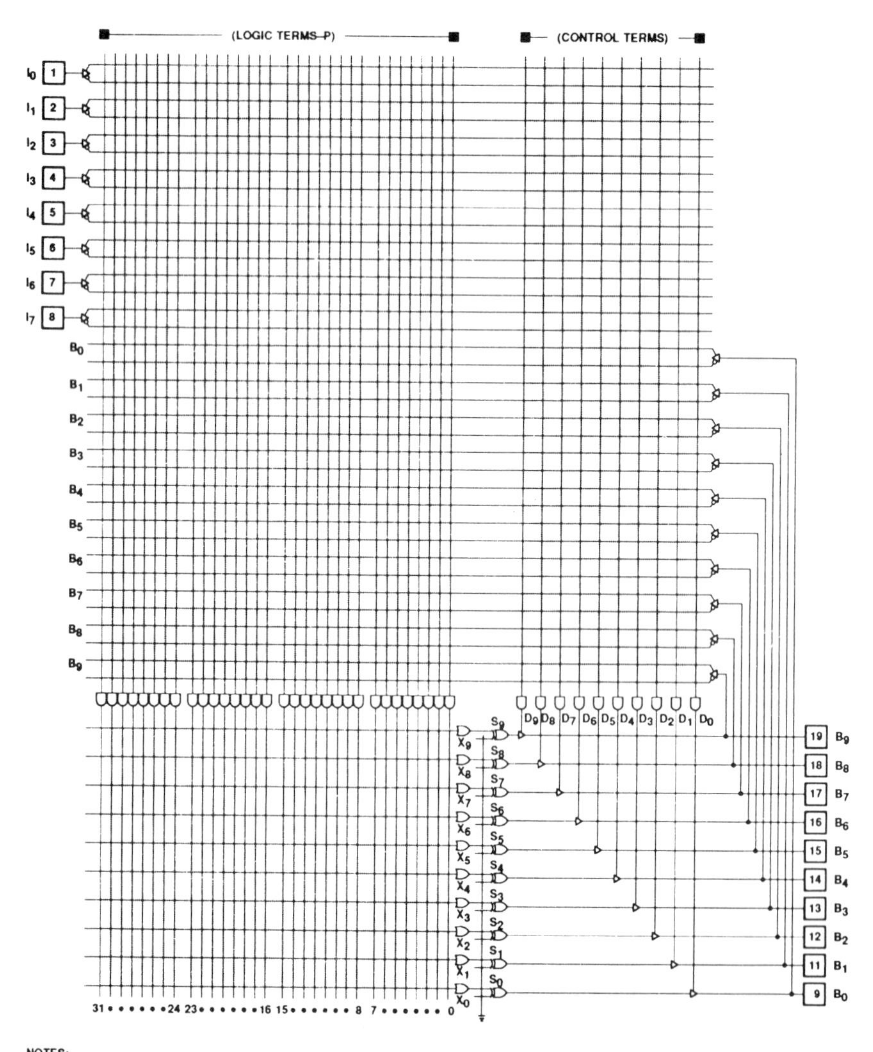

Figure 3-15
Signetics PLS-153. (Reprinted courtesy of Philips Components-Signetics, ©1990.)

example, things like a 16 input, 2 output architecture. This can be very useful for monitoring illegal states on 16 bit busses, etc. Since the output buffers are under control of the product terms, the I/O pins can even be used for output functions part of the time and inputs the rest of the time.

3.4 A Typical PLD Application

As noted in our discussion of the PROM, one of the most common uses of PLDs is in address decoding. To demonstrate a typical example, we will look at a microprocessor system. A block diagram of our system is shown in Figure 3-16. The system uses the Motorola 6809 microprocessor, industry standard memory, and peripheral devices. The memory map for this system is shown in Figure 3-17. Our task is to realize the block marked "Decoding Block" in Figure 3-16. Two things in this design scheme are of particular interest. First is the interface timing of the 6809 and second is the address mapping of the 32K RAM chip.

The timing of the 6809, as is the case for most Motorola processors, is synchronous. This means that for a read or a write to occur, a single read/write (RW) line is used. The peripheral being addressed must look at a second line, called the E line, to know when the read/write is active. Conventional memory devices, however, make use of an asynchronous read line and an asynchronous write line. Our address decoding block

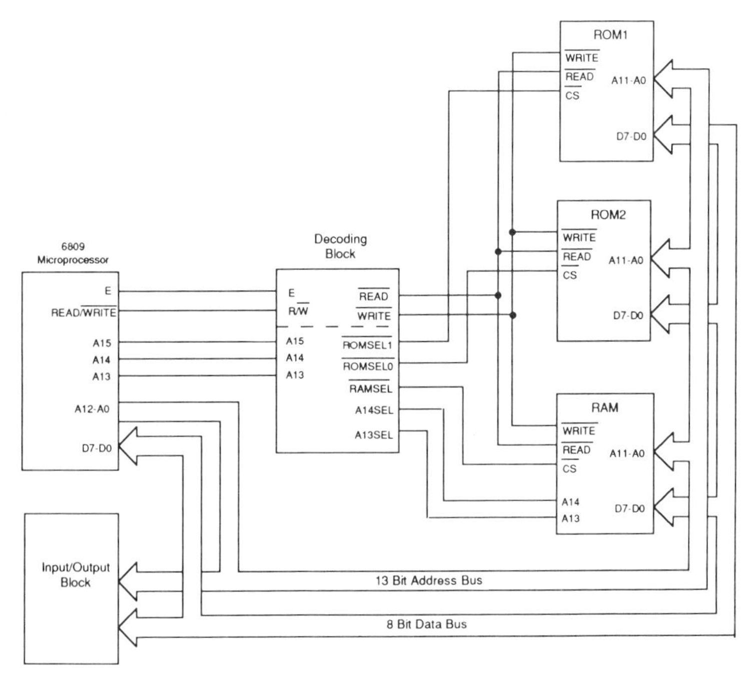

Figure 3-16
6809 Address Decoder Example.

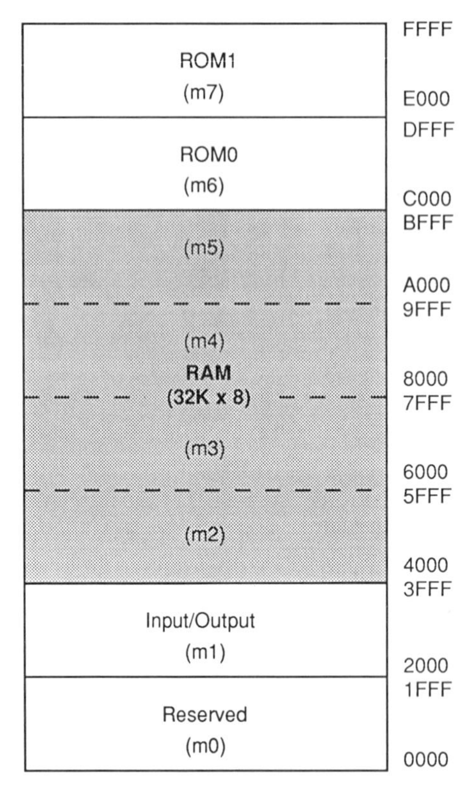

Figure 3-17
Address Map for the 6809 Example.

must convert the 6809's synchronous timing to the asynchronous timing of the memories.

The RAM also presents an interesting problem. We have chosen a standard 32K $\times$ 8 bit RAM chip. This would not be a problem if we could put it in either the upper or lower half of the 6809's 64K address space. As it works out, however, we must place it in the middle of the address space. This means that the upper address bits of the RAM cannot be used as address bits, since they would not correctly map into the decoded address. As shown in Figure 3-16, the solution to this problem is to treat the upper two address bits of the RAM as additional chip selects.

Now, let us see how we go about realizing the decoding block. The read/write timing is easy if we consider that the /READ signal is simply the status of the RW line ANDed with the E signal, then inverted. Mathematically:

$$\overline{\text{READ}} = \overline{(\text{RW} * \text{E})} \qquad (3\text{-}2)$$

The /WRITE timing signal is simply the INVERTED RW line ANDed with the E signal:

$$\overline{\text{WRITE}} = (\overline{\text{RW}} * \text{E}) \tag{3-3}$$

A truth table is shown in Figure 3-18a.

Figuring out the decoding is more interesting. There are a number of ways of approaching the problem. Simple inspection is usually the fastest. For this discussion we will take a slightly more formal and therefore less heuristic approach.

Input		Output	
E	$R/\overline{W}$	READ	WRITE
0	0	1	1
0	1	1	1
1	1	0	1
1	0	1	0

Read/Write Decoding
(a)

	Input			Output					
Minterm	A15	A14	A13	ROM1	ROM0	RAMSEL	RAM14	RAM13	IOSEL
m7	1	1	1	0	1	1	X	X	1
m6	1	1	0	1	0	1	X	X	1
m5	1	0	1	1	1	0	1	1	1
m4	1	0	0	1	1	0	1	0	1
m3	0	1	1	1	1	0	0	1	1
m2	0	1	0	1	1	0	0	0	1
m1	0	0	1	1	1	1	X	X	0
m0	0	0	0	1	1	1	X	X	1

Address Decoding
(b)

Figure 3-18
Decoding Table for the 6809 Example.

The first thing to do is to break the address space into two parts: The upper bits become the decoding field and the lower bits become the byte select field. To determine how many bits are in each field, we first determine the granularity of the decoding. From Figure 3-15, we can see that the smallest range of addresses that must be selected is 8K. Dividing the 64K address by the 8K range gives us eight partitions. The number of bits required to encode the eight partitions is simply $\log_2(8) = 3$ bits. Thus, the three high-order address bits A15, A14, and A13 must be used for decoding. A12 through A0 are used for byte selection.

Now that we know how the decoding layout in Figure 3-15 was determined, let us look at how we can arrive at the basic equations for realizing our circuit. Figure 3-18b is a table of the requirements. Notice that in tabulating the range of address for A15, A14, and A13, we have simply generated all of the minterms for these three variables. For example:

$$\begin{aligned} \overline{\mathrm{ROMSEL1}} &= m_7 \\ &= \mathrm{A15} * \mathrm{A14} * \mathrm{A13} \end{aligned} \tag{3-4}$$

Similarly

$$\begin{aligned} \overline{\mathrm{ROMSEL0}} &= m_6 \\ &= \mathrm{A15} * \mathrm{A14} * \overline{\mathrm{A13}} \end{aligned} \tag{3-5}$$

and

$$\begin{aligned} \overline{\mathrm{IOSEL}} &= m_1 \\ &= \overline{\mathrm{A15}} * \overline{\mathrm{A13}} * \mathrm{A13} \end{aligned} \tag{3-6}$$

We need not worry about the "RESERVED" address space, but it would decode to m_0.

The interesting discussion is, of course, the decoding for the RAM. We want the $\overline{\mathrm{RAMSEL}}$ signal to be low whenever any of the four partitions covered by the address range of the chip are encountered. In other words:

$$\begin{aligned} \overline{\mathrm{RAMSEL}} &= m_2 + m_3 + m_4 + m_5 \\ &= (\overline{\mathrm{A15}} * \mathrm{A14} * \mathrm{A13}) + (\overline{\mathrm{A15}} * \mathrm{A14} * \mathrm{A13}) \\ &\quad + (\mathrm{A15} * \mathrm{A14} * \mathrm{A13}) + (\mathrm{A15} * \overline{\mathrm{A14}} * \mathrm{A13}) \end{aligned} \tag{3-7}$$

We could do the same for the RAM's address lines being used as chip selects: RAM14 and RAM13. However, if we look at the table in Figure 3-18b, we notice some simplifying facts. Notice that RAM13 happens to have exactly the same value as A13. Thus, while we still need A13 as a decoding input, we can use A13 to directly drive RAM13. We could do this by simply connecting A13 on the RAM chip to A13 on the 6809. Or to keep with the design layout we could simply say:

$$RAM13 = A13 \tag{3-8}$$

For RAM14 the situation is not as simple, but notice that RAM14 is just A14 INVERTED:

$$RAM14 = /A14 \tag{3-9}$$

We now have all the information needed to fabricate the decoding circuit. The decoder could be implemented with either a PROM, a PAL, or a PLA. For a PROM, we could develop a table such as the one shown in Figure 3-4, or we could take Eqs. (3-4) through (3-9) and feed them into an assembler or compiler for a PAL or PLA.

Alternatively, for a compiler we could simply use a table similar to Figure 3-18. In Chapter 6 we will see in greater detail how to do this.

3.5 PROM, PAL, and PLA Comparisons

Now that we have discussed the three basic PLD architectures, the PROM, the PAL, and the PLA, it is useful to make some comparisons and contrasts. We do this by describing each of the three devices in a common PLD form. The three forms are shown in Figure 3-19.

As can be seen in Figure 3-19a, the PROM is basically a fixed AND array, programmable OR array device. This is necessary in a memory device where every possible minterm must be generated. Each minterm selects the address of the word being addressed, and the output word is the selected ORing of each minterm.

As noted earlier, however, most logic functions require only a small subset of the possible minterms. This means that PROMs are inherently inefficient as PLDs. This inefficiency is further worsened by the architecture of the PROM. In its role as a memory device, it must have fixed input and fixed output pins. But for a PLD it is more useful to have as much flexibility for assigning inputs and outputs as needed.

This type of flexibility is demonstrated in the PLS153A where it can have as many as 17 inputs and 0 outputs [a classic write only memory (WOM)] to a more balanced 8 inputs and 9 outputs.

These factors limit the typical role of the PROM in programmable logic implementations to applications that are not too dissimilar from the PROM's classic use as a memory device. These applications include state machines and microsequencers, which we will be discussing in later chapters.

To overcome the limitations of the PROM, Signetics developed the versatile and flexible PLA (Figure 3-19c). The PLA fits the needs of the logic designer well and is theoretically superior to the PAL. This leads to

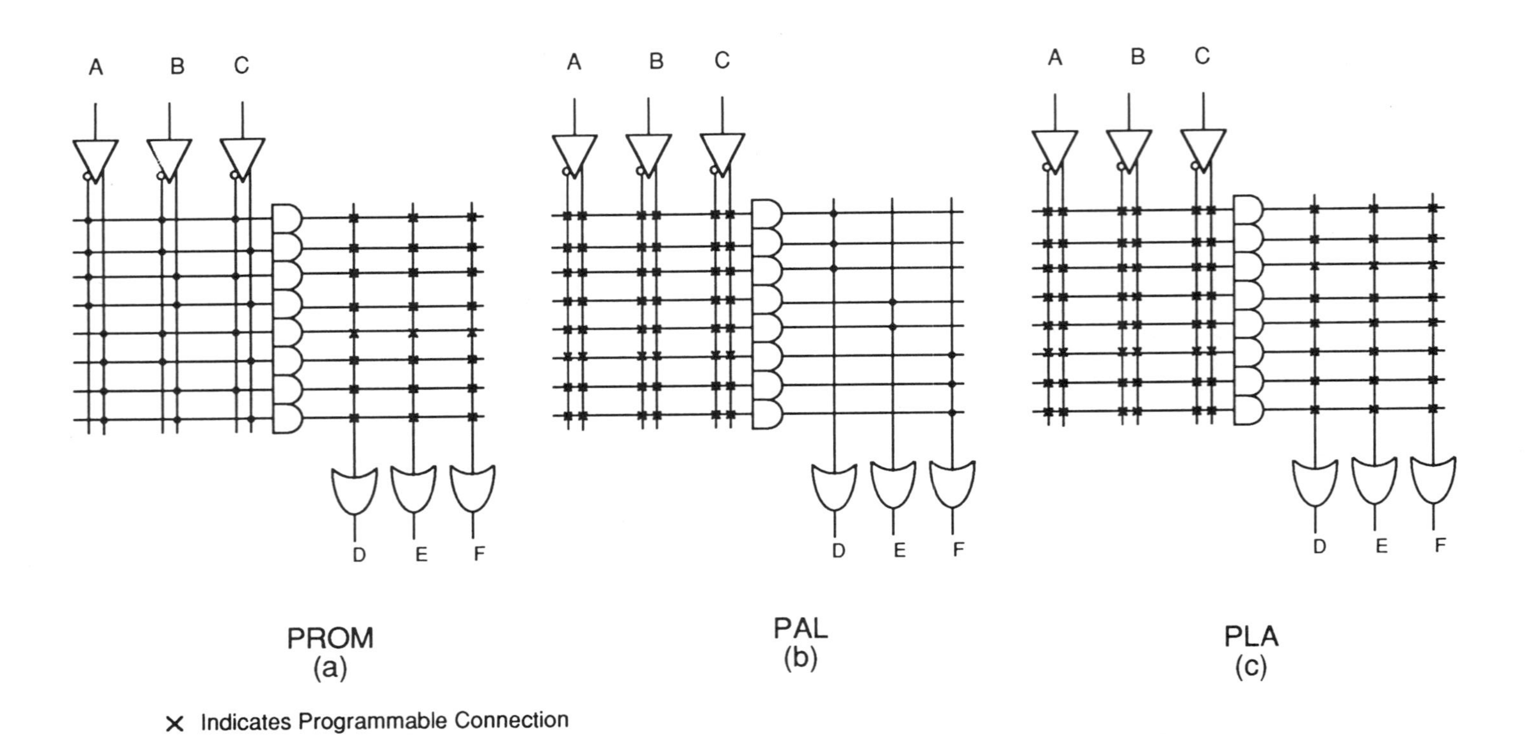

Figure 3-19
PLD Architecture Comparisons.

the interesting question of why the apparently backward step of introducing the far more popular PAL was required.

There are several answers. First are simple market factors. The PLA led to the introduction of PLDs. There was a certain learning curve that had to be overcome to get designers accustomed to using PLDs. Another market factor was the lack of user friendly software to support the use of PLAs. This was by no means a limitation unique to Signetics; most chip manufacturers in the mid-1970s felt that their role was limited to producing silicon. Software was something that came from some other source. The first PLAs were relatively slow and costly. The two programmable matrixes were difficult to fabricate, thus making production yields low. The two arrays also made for relatively long propagation times through the PLA.

These factors set the stage for the introduction by MMI of the simpler, more usable PAL (see Figure 3-19b). The fixed OR architecture of the PAL made it easier to program and easier to write supporting software for programming. Eliminating the programmable OR array improved yields and reduced propagation delays.

While it was true that the PAL was more limited than the PLA, this fact was offset by two things. First, a wide variety of PALs were introduced. Second was the question of whether the extra power of the PLA was actually required. Today it is generally acknowledged that the PAL architecture is completely adequate for most functions. Since it is cheaper and simpler than the PLA, it is generally the logical choice. The more sophisticated PLA architecture is reserved for special applications requiring greater flexibility.

The actual devices discussed in this chapter, the PAL16L8 and PLS-153, while by no means obsolete, are mature designs. We have discussed them for a variety of reasons. First, of course, is the fact that the engineer working with PLDs will commonly encounter these types of devices. This is particularly true in designs that are a few years old. Second, studying these "first generation devices" makes it easier to understand the second and third generation devices we will be discussing in the upcoming chapters.

3.6 Chapter Summary

- A PROM can be used to store any logic function as long as the number of variables is less than or equal to the number of address lines on the PROM. This is analogous to saying that a PROM can address any word in its program space.

- For a PROM, the number of logical outputs is equal to the number of data lines (usually eight).
- Architecturally, a PROM is a fixed AND array, programmable OR array device.
- The number of product terms in a PROM is equal to 2^n, where n is the number of inputs.
- A PAL is a programmable AND array, fixed OR array device.
- The PAL architecture is well suited to most but not all logic equations.
- The percentage of resources used per function in a PAL is a function of its architecture, *not* of the complexity of the function.
- The PAL architecture is the most popular PLD architecture.
- The PLA is a programmable AND array, programmable OR array device.
- The PLA is the most flexible of all combinatorial PLDs. One usually pays for this in both the speed and cost of the device.
- Like the PROM, the PLA can handle any logic function within its range. It is more efficient than the PROM, however, since the same product term can be used by several OR terms.
- PAL-type devices work best when a large number of inputs are required, and each output is relatively independent of other functions in the PAL. Applications requiring sparse usage of the AND array and limited use of the (fixed) OR array are the best fit. Glue logic often makes use of PALs.
- PLAs are the best choice when a large number of inputs are required but there is more commonality between the outputs. PLA applications also typically make sparse use of the AND array, but they generally make more use of the OR array.
- PROMs are best when a limited number of inputs are required but there is a strong interdependence among the outputs. PROM applications are characterized by heavy use of both the AND array and the OR array. Typical applications include arbitrary waveform generators, transforms and translation tables, etc.

4 State Machines

Up to this point, basic combinatorial and sequential logic operations have been discussed. The basic architectures used by PLDs to implement combinatorial circuits have also been presented. Before we can meaningfully discuss the more sophisticated PLDs that are now available, we must cover two additional topics. In this chapter we will cover the first topic, state machine design. The second topic is software development tools, introduced in this chapter and more fully developed in Chapter 5.

Most advanced PLDs are designed with an orientation toward realizing state machines. Even when these PLDs are not used as true state machines, they often make use of state machine like latches, counters, and combinatorial logic.

This chapter will present a basic introduction to the theory of state machines. We will cover the basic architectures of the classic Mealy and Moore state machines. Two practical examples showing how PROMs and latches are used in state machines will be demonstrated. In the process, the field of microcoding is introduced.

With an understanding of state machines and software tools to work with, Chapter 6 will look at the more sophisticated PLDs that are available to the modern design engineer.

4.1 An Introduction to State Machines

State machine circuits are a major part of many modern electronic circuits. They are found in VLSI components such as UARTs, error detection and correction chips (EDACs), and central processor units

(CPUs). State machines are at the core of circuits such as sequencers, arbitrary waveform generators, and bus arbitration circuitry.

From a programmable logic point of view, state machines are on a line between combinatorial PLDs and microcontrollers. State machines are sophisticated enough to handle sequential situations where simple decisions are required. On the other hand, their simple architecture makes them much faster than microcontrollers.

Technically, almost any combinatorial circuit with some feedback can be thought of as a state machine. In practice, however, the term is generally reserved for synchronous state machines that follow one of the two major architectures, Mealy or Moore. These architectures will be discussed shortly. In addition to the two basic architectures, state machines can be further characterized as either synchronous or asynchronous.

Asynchronous state machine design is analogous to our asynchronous counter described in Chapter 1. The clocks of the individual latches are driven separately. These designs have the advantage of being simpler in terms of gate count than are synchronous designs. This simplicity makes them desirable for applications where speed is paramount or where the maximum logical power must be fit into the minimum possible space. Asynchronous designs are much harder to validate in terms of timing, are more difficult to design, and are much harder to modify successfully. For these reasons, asynchronous designs are relatively rare, especially for programmable logic designs. In this chapter, the term *state machine* refers to synchronous designs of either the Mealy or Moore architecture.

Synchronous state machine design methodologies developed as a means of efficiently dealing with the growing complexity of sequential circuit design. As we pointed out in Chapter 2, it is easy to get into situations in which glitches in sequential logic occur. While good design practice will keep these problems under control in simple circuits, it becomes a major problem as sequential circuits become more complex. Even more daunting is the task of modifying complicated sequential designs. As any engineer who has tried modifying a complicated design of discrete flip-flops and gates can attest, it becomes virtually impossible to successfully modify such a design. It is often easier to redesign the circuit from scratch.

An analogy can be made between state machine design to the hardware engineer and structured programming for the software engineer. Networks of flips-flops and gates are analogous to in-line code with many "goto" statements. State machines are analogous to structured code where tasks are broken out into well-defined functions. It is this analogy that often sparks the characterization of state machines as "hardware subroutines."

The modularity of design possible with state machines is particularly important in CPU designs. It is a major task to get a CPU working with hard-wired logic. Once a design is working and validated, changes become very expensive. Adding a new instruction or modifying one that has been found to have a subtle bug is prohibitively expensive. The use of state machines in the design of the CPU can make these types of changes simple and very cost-effective. We will look at this further in Chapter 8.

4.2 Classic State Machines

The classical implementation of the state machine is with a simple PROM and a data latch. This organization is conceptually the simplest and is the one we will discuss in detail in this chapter. In later chapters we will look at sophisticated PLDs that make very powerful state machines. The classic PROM/latch implementation retains many advantages, however. A wider number of conditional inputs, an unlimited number of outputs, and a far greater number of states are possible with the classic architecture. Furthermore, a classically implemented state machine is easier to debug since all of the key signals are accessible on the circuit board.

Figure 4-1 shows the two basic state machine architectures, the Mealy and Moore. Mealy state machines are characterized by the fact that their output is a function of both the *current state* and the *conditional input*. The output of the Moore state machine, on the other hand, is only a function of the current state. We will look at these definitions in greater detail in a moment.

The block marked "Combinatorial Logic" in Figure 4-1 is typically a PROM or PLD. The latch is typically a high-speed TTL component such as a 74S273 or a 74HC273. If we think of the combinatorial logic as a PROM, then the inputs on the left-hand side of the combinatorial block are the address lines and the outputs on the right-hand side are the data lines.

The clock input to the latch periodically updates the state of the overall machine. The reset line simply clears all of the D-registers. The D-registers are often called *pipeline* registers. The term derives from the use of these registers in pipelined microsequencer applications. Pipelining is discussed in Chapter 8.

Up to this point, we have been discussing state machines in essentially textbook terms. It is easier to understand exactly what state machines do and why they are useful in real world circuits by examining some practical applications.

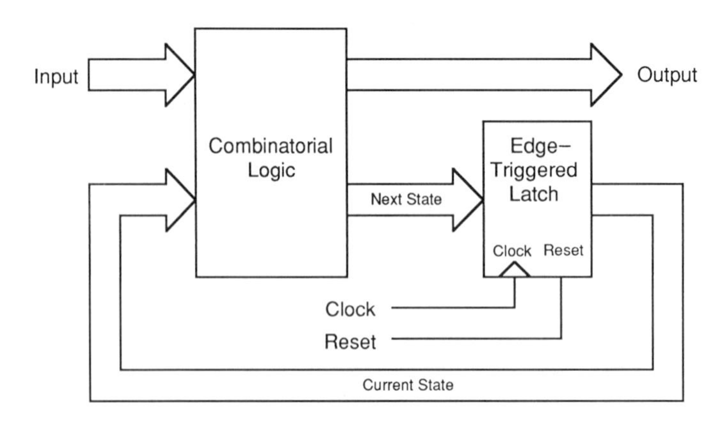

Mealy State Machine Model
(a)

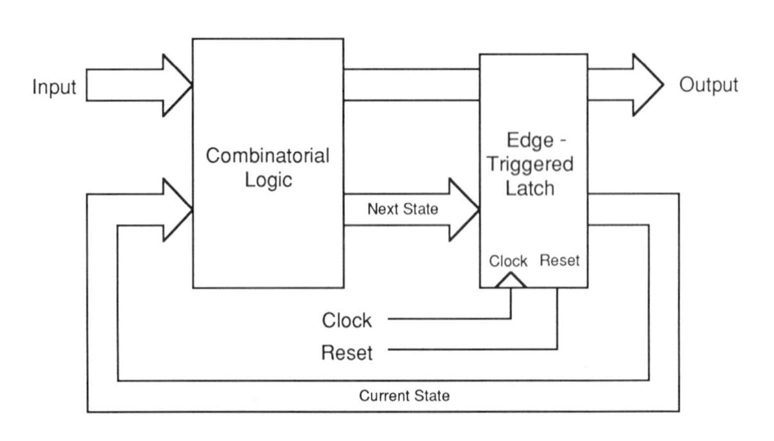

Moore State Machine Model
(b)

Figure 4-1
Two Basic State Machine Models.

For the first example we will return to the basic counter circuit from Chapter 2. In Chapter 3 we saw how a PROM or PLD could be used as a display decoder. In this chapter, we will look at how to replace the heart of the counter, the J-K flip-flops, with a PROM and an edge-triggered latch.

J-K flip-flops are simple and inexpensive so the first question is "why would we want to replace them with a relatively complex and expensive PROM and edge-triggered latch?" In part, of course, we are doing it to have a simple, easily understood example for our state machine discussion. But there are several additional advantages. As we will see, by using a state machine the circuit gains some useful attributes:

- We can program the outputs to be whatever we want. This is useful for things such as Gray counters, Johnson counters, or other non-standard counting sequences.
- We gain greater flexibility. We will add, for example, the capability to count down as well as count up.
- While it is not a problem with our simple counter, the J-K-based design is asynchronous. Our state machine design will inherently be synchronous. This improves testability and overall performance and makes for a more rugged design.
- The state machine design is flexible. If we decide we want a base 10 counter, a base 16 counter, or any other base, it is simply a matter of changing the PROM. No circuit redesign or rework is required.

To get a feel for how a state machine really works let us look at the actual circuit.

Figure 4-2 is a schematic of our state machine–based 4 bit counter. Notice the close similarity of the actual circuit to the theoretical block diagram of Figure 4-1a. The RESET line of the D-registers is connected to

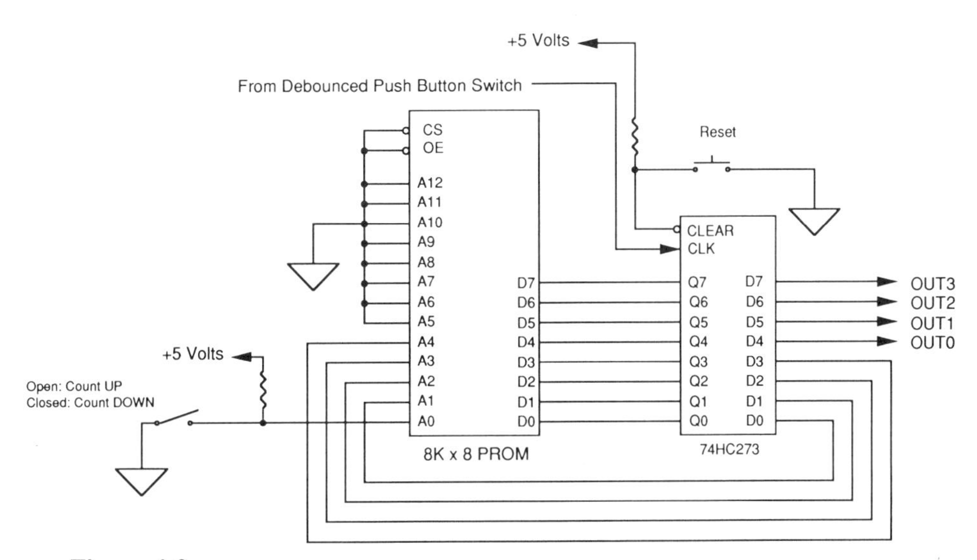

Figure 4-2
Moore State Machine Example (Four Bit Counter).

the RESET button. Pressing the reset button causes all of the latches to clear (i.e., go to 0). The clock comes from our debounced switch shown in Figure 2-12. A debounced switch is necessary to insure that states change only once for each push of the button.

Notice the way the feedback lines forming the current state are connected to the address inputs of the PROM. The state lines are connected to the high-order address bits. The conditional input, in this case the up/down select switch, is connected to the low-order input. This may seem counter-intuitive (it always has to the author, anyway). The reason for it will be made clear shortly.

In this case we have only the one conditional input, the up/down select line just mentioned. We could have used the other unused address lines to afford extra functionality to the circuit. For purposes of illustration, however, one conditional input is adequate.

The output of the state machine is simply sent directly to the display driver. Since we are making use of four bits of feedback, our counter can count as high as $2^4 = 16$ (0 through 15). To display all 16 counts we would need the hexadecimal display driver from Chapter 3. If we had restricted the counter to decimal counting, we could have used the 74HC4511 from Figure 2-16.

Figure 4-3 shows a state diagram of the counter. The arrows simply show which is the next state. If the conditional input A0 is a 1, the counter increments. If A0 is a 0, the counter decrements.

Now that we have seen how the circuit is implemented and what it actually does, let us look at how it is programmed. This programming is best illustrated by looking at a *state table*. Such a table is shown in Figure 4-4. The left-hand column shows each state that our counter can assume. Notice that for each state there are two addresses. If n equals the number

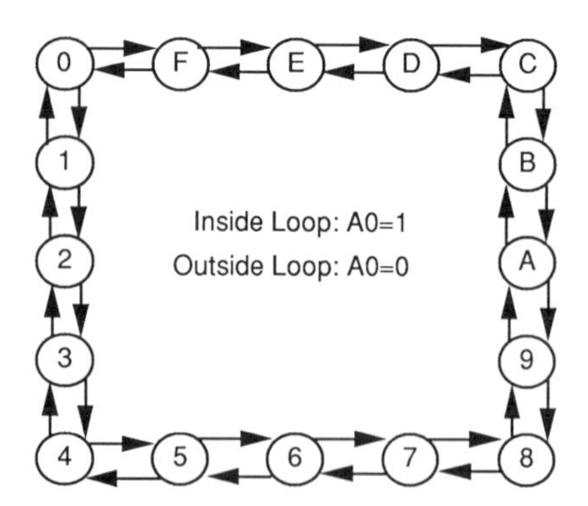

Figure 4-3
State Diagram for a Four Bit Counter.

Up/Down Conditional Input.

	Address					Data							
	Currernt State					Output				Next State			
State	A4	A3	A2	A1	A0	D7	D6	D5	D4	D3	D2	D1	D0
0	0	0	0	0	0	1	1	1	1	1	1	1	1
	0	0	0	0	1	0	0	0	1	0	0	0	1
1	0	0	0	1	0	0	0	0	0	0	0	0	0
	0	0	0	1	1	0	0	1	0	0	0	1	0
2	0	0	1	0	0	0	0	0	1	0	0	0	1
	0	0	1	0	1	0	0	1	1	0	0	1	1
3	0	0	1	1	0	0	0	1	0	0	0	1	0
	0	0	1	1	1	0	1	0	0	0	1	0	0
4	0	1	0	0	0	0	0	1	1	0	0	1	1
	0	1	0	0	1	0	1	0	1	0	1	0	1
5	0	1	0	1	0	0	1	0	0	0	1	0	0
	0	1	0	1	1	0	1	1	0	0	1	1	0
6	0	1	1	0	0	0	1	0	1	0	1	0	0
	0	1	1	0	1	0	1	1	1	0	1	1	1
7	0	1	1	1	0	0	1	1	0	0	1	1	0
	0	1	1	1	1	1	0	0	0	1	0	0	0
8	1	0	0	0	0	0	1	1	1	0	1	1	1
	1	0	0	0	1	1	0	0	1	1	0	0	1
9	1	0	0	1	0	1	0	0	0	1	0	0	0
	1	0	0	1	1	1	0	1	0	1	0	1	0
A	1	0	1	0	0	1	0	0	1	1	0	0	1
	1	0	1	0	1	1	0	1	1	1	0	1	1
B	1	0	1	1	0	1	0	1	0	1	0	1	0
	1	0	1	1	1	1	1	0	0	1	1	0	0
C	1	1	0	0	0	1	0	1	1	1	0	1	1
	1	1	0	0	1	1	1	0	1	1	0	1	1
D	1	1	0	1	0	1	1	0	0	1	1	0	0
	1	1	0	1	1	1	1	1	0	1	1	1	0
E	1	1	1	0	0	1	1	0	1	1	1	0	1
	1	1	1	0	1	1	1	1	1	1	1	1	1
F	1	1	1	1	0	1	1	1	0	1	1	1	0
	1	1	1	1	1	0	0	0	0	0	0	0	0

Figure 4-4
State Table for a Four Bit Counter.

of conditional inputs, each state will be composed of $2^n = 2^1 = 2$ addresses. The next major column shows the address inputs into the PROM. The address column is subdivided by the dashed line into the current state lines and the conditional input line. The combination of the current state and the conditional inputs form the *effective address* for the PROM. The final major column is the data byte stored in the PROM. The upper nibble is the logic output. The lower nibble forms the next state.

Each byte in our programming table is known as a *microword*. The process of generating the data to be programmed into the PROM is called *microcoding*. The prefix "micro" is of historical interest. The term derives from the use of special state machines, known as microsequencers, used in the design of CPUs. To differentiate this firmware coding from the user-generated assembly language programming, the prefix "micro" is attached. This topic is developed further in later chapters.

It is easiest to understand the operation of a state machine by following the operation through a few cycles. Referring to Figure 4-3 and Figure 4-4 may help to provide mental reference points during the discussion.

Let us assume the up/down switch is open. This will place a 1 on the conditional input. Now assume that we push and release the reset button. All of the flip-flops in the 74HC273 will go to 0. The outputs, OUT0 through OUT3, will indicate zero for the count. The current state, that is the state information at the output of the 74HC273, is of course 0000. The combination of the current state and the conditional input form an effective address of 0 0001 into the PROM. If we look at the address 0 0001 in the table, we see that the output is 0001 and the next state is 0001. These are the values that are on the input of 74HC273. And so our state machine sits until it is clocked.

When a clock pulse is seen by the 74HC273, the output value of 0001 and the next state value of 0001 will be clocked to the *Q* outputs of the latch. We will have counted from 0 to 1. If we leave A0 high, the effective address to the PROM will now be 0 0011 (i.e., state 2). The next clock would send us to state 2. As long as A0 remains high, the counter will continue to increment. If we set A0 to 0, however, the effective address would be 0 0010. This would place 0000 0000 on the input to the D-register. The next clock would transition back to state 0.

At the limits, our counter is set up so that it underflows from 0 to *F*. Likewise, it overflows from *F* to 0.

This is a simple example, but it contains all of the key elements of state machine design. It is a good mental exercise to follow the clocking of this circuit through various counts. More sophisticated and complex designs will be encountered often in the real world. However, if one understands this circuit, all the others are just variations on the theme.

Some further observations on the programming of this counter circuit are in order. In this case we sequentially stepped through each state, going up or down one state as desired. Outside the fact that this is the way a counter should behave, we were not restricted to this sequence. If for some reason we wanted to follow the sequence 0-7-*A*-3-*F*, for example, we could have. All we would have had to do is program the next state values appropriately.

The outputs simply replicated the state information. That is, if the counter is in state 0, the output is a 0. In state 1, the output is a 1. This is not necessarily the case in all designs. We could make the outputs whatever we wanted them to be.

In fact, if there were such a thing as an 11 bit PROM and an 11 bit version of the 74HC273, we could have used seven outputs. These seven outputs could have driven the seven-segment display directly. While there is no such thing as an 11 bit wide PROM commonly available, we will see in a later chapter how advanced PLDs allow us to accomplish essentially the same thing.

Another interesting thing to notice is the effect of the conditional inputs on the output of the state machine. Notice that changing the value on A0 has no immediate effect on the output. It is only after a clock pulse has been received that the PROM's outputs are clocked to the output of the state machine. For this reason, we say that the output is only a function of the current state. This makes the state machine a Moore architecture.

If we had taken the outputs directly from the PROM rather than through the 74HC273, we would have had a Mealy machine. In the case of our counter, this would not have made a theoretical difference; the output from the PROM is the same for either a 0 or a 1 on A0. From a practical point of view, however, there is some advantage to the Moore design. Changing A0 could cause brief glitches on the output while the PROM's address settled. This potential problem is eliminated with the Moore implementation.

Since the output of a Mealy state machine is immediately affected by the conditional input, we say that the output is a function of both the current state *and* the conditional input.

State machine design techniques are often difficult to master initially. This typically results from the lack of intuitively obvious guideposts in the design. The conditional inputs, as noted, are often the lower order inputs. This is done not for any hardware reason but rather because it simplifies the layout of the programming table shown in Figure 4-4. By making the conditional input the low-order address, we can keep the individual addresses associated with each state together. As we are about to see, this is handy when we use software development tools for building the table.

The process of laying out the bit assignments is called *defining the microword* (Figure 4-4). This is a critical step in the design process, since once the microword has been defined, the actual coding will tend to fall into place. As we saw with the previous example, the mechanics of state machine operation are actually quite simple if one takes the time to mentally follow the flow.

For the 4 bit counter example just shown, directly generating the microcode by filling in the programming table was not only feasible, it was also desirable. The objective was to provide a strong intuitive base to show how the code actually works in the circuit. For more complex designs, however, it is handy to add some abstraction to the microcoding process.

For example, we would like the ability to name the states symbolically. This frees us (some) from worrying too much about the individual addresses. A more abstract representation also allows the addition of comments and other descriptors to our code. Having the code symbolically represented in a source file also makes the process of maintaining and updating the code a simpler procedure.

Once all of the microcoding information has been put into an ASCII source file, we can use a tool known as a microcode assembler (also commonly called a meta-assembler) to translate the source file to an object file. To illustrate the techniques we will look at another application of a state machine, an electronic combination lock.

First, however, let us look back at our counter circuit. The counter was clocked by the debounced push button. The state of the machine was updated for each button press. Thus each state change was nonperiodic. In some designs the clock input is not synchronously controlled like this. For example, in timing circuits we may want to apply a free-running clock to the state machine. This is often the case for state machines used as dynamic memory controllers or for bus arbitration circuitry. The free-running clock sets the overall timing base and the conditional inputs simply select which state is to be entered next.

To illustrate how this works, the next example will make use of a free-running clock. Figure 4-5 shows the schematic of an electronic combination lock. The circuit is simplified inasmuch as it has only three buttons. Typically, this type of lock has five buttons, but to make the code more understandable we have simplified things a bit.

A state diagram for the lock is shown in Figure 4-6. The lock's operation is very straightforward. To open the lock the three buttons must be pushed in the correct sequence. For the combination of 2-3-1, we must first push button 2, release it, push button 3, release it, and finally push

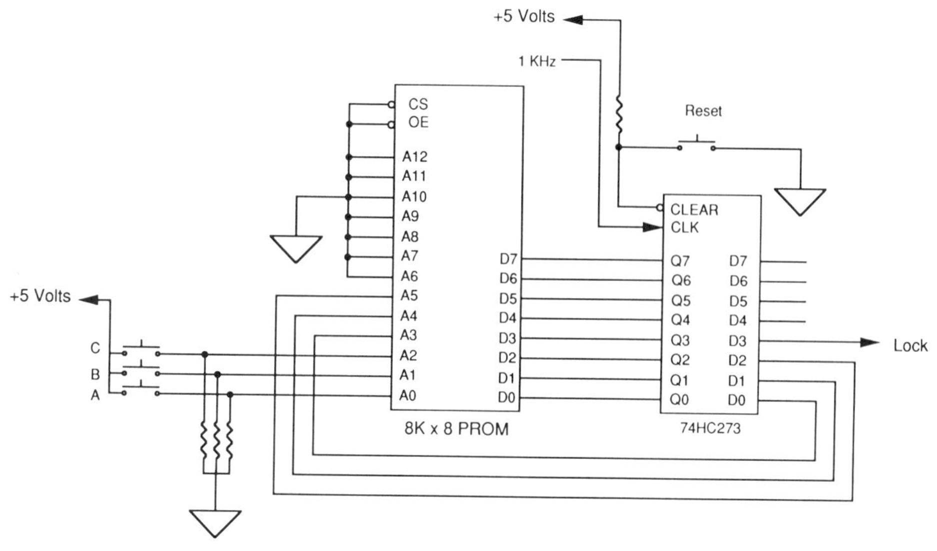

Figure 4-5
Combination Lock Schematic.

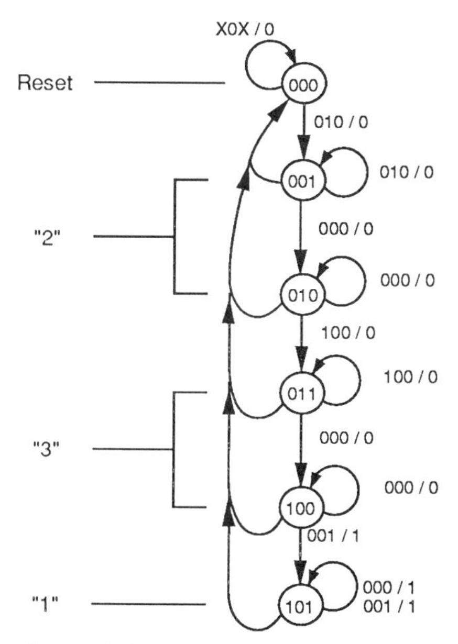

Figure 4-6
State Diagram for the Combination Lock.

button 1. When we push button 1, the lock will open. It will remain open until we push any button other than button 1.

From this description we know enough information to specify the size of our microword. As noted previously, the microword size is equal to $k + n$, where k is the number of bits required to encode all of the states and n is the number of conditional input terms. In the current example there are three switches, so $n = 3$.

As shown in Figure 4-6, we need six states to implement the combination. Two bits would encode four states, which are not enough. Three will encode eight states, so $k = 3$. Or formally:

$$\begin{aligned} k &= \text{int}(\log_2(6) + 1) \\ &= 3 \end{aligned}$$

The microword must have three input bits and three state bits for a total of six inputs. The outputs are simply $k + m$, where m is the number of output bits used for purposes other than state information. Therefore, we need four output bits.

To hold every possible state, our PROM must then be at least $2^{(3+3)} = 64$ bytes.

It used to be possible to buy things like 64 × 4 bit PROMs, but they are becoming very hard to find. To keep things simple, we have used the ubiquitous 8K × 8 EPROM and simply grounded the unused address lines. The high-order nibble of the PROM's output is simply left floating.

In this example, the clock frequency is not particularly important as long as it is short in comparison to human perception time. A typical value would be between a few hundred and several thousand hertz.

Referring once again to Figure 4-6, note that state 000 is the reset state and the state to which any invalid push of a button transfers to. The arrow looping back on state 000 shows that any condition other than the exiting condition causes the state machine to stay in the 000 state. The exiting condition is the pushing of button B. This condition is shown in the figure as an arrow from state 000 to state 001. The notation next to the arrow, 010/0, signifies two things. The 010 to the left of the "/" is the input that causes the transition. The 0 to the right of the "/" is the output of the state machine.

State 000 is the state immediately after the D-register is cleared. The lock is closed. If no buttons are pushed, the state machine simply loops in state 000. Remember, this is also the target state of any transition that is a wrong sequence in the combination.

As noted, state 001 is reached by pushing button B (010) while in state 000. As long as the button is pushed down, the state machine will loop in state 001. When button B is released (000) the state machine will transition to state 010. We will stay in this state until another button is pushed. Two states are required for each digit in the combination: one for the period while the correct button is depressed and one for when the button is released.

If the next button in the combination sequence is pressed, button C, we will transition to state 011. If any other button is pushed, we will transition to state 000.

This sequence follows through state 100 and into state 101. Notice that on entering state 101, the LOCK output changes from a 0 (the lock is closed) to a 1 (the lock is open). In state 101, we simply loop for either a 000 conditional input (no buttons are pressed) or for 001 (button A remains pressed). The lock is closed by pressing any button other than button A.

State diagrams are a simple and efficient way of describing the operation of a state machine. They do not, however, lend themselves well to the actual code generation process. Instead we will make use of a textual definition for our design.

To help the visualization process, a programming table for the lock example can be defined. A portion of the table is shown in Figure 4-7. This is not technically necessary since we will be generating this information in our microcode source file. The programming table is included, however, as an aid to visualizing the microcode generation process.

The first thing that we must do is define the instructions that we want the state machine to follow. Like everything else so far, the microinstructions get the "micro" prefix to differentiate them from normal computer instructions.

For this example, we will define three microinstructions: CLEAR, STEP, and OPEN. We could have defined more or fewer microinstructions. There is nothing critical about the three we picked, they are just a convenient partitioning.

We will define the microinstructions in two ways. First, we will define them heuristically. Then we will define them formally so our meta-assembler can make use of them.

Heuristically, the three microinstructions are defined as:

CLEAR This instruction does two things. It defines the lock to be closed and the next state transition to be state 0. This will be the most commonly used microinstruction.

STEP This is the microinstruction for stepping from one correct state in the sequence to the next state in the sequence.

OPEN This sets the output to 1, thus opening the lock.

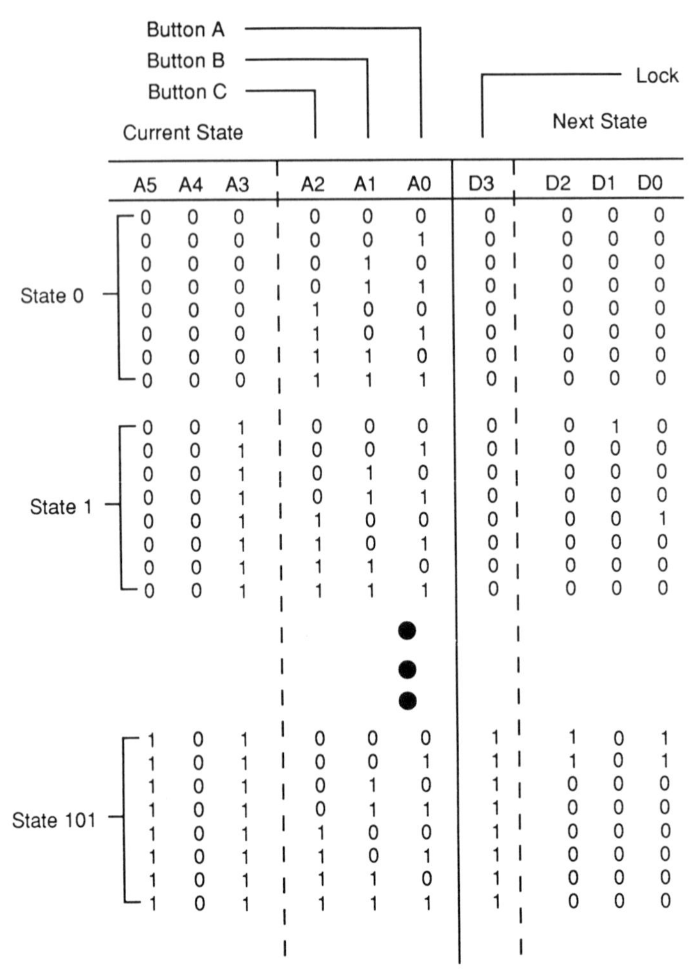

	A5	A4	A3	A2	A1	A0	D3	D2	D1	D0
State 0	0	0	0	0	0	0	0	0	0	0
	0	0	0	0	0	1	0	0	0	0
	0	0	0	0	1	0	0	0	0	0
	0	0	0	0	1	1	0	0	0	0
	0	0	0	1	0	0	0	0	0	0
	0	0	0	1	0	1	0	0	0	0
	0	0	0	1	1	0	0	0	0	0
	0	0	0	1	1	1	0	0	0	0
State 1	0	0	1	0	0	0	0	0	1	0
	0	0	1	0	0	1	0	0	0	0
	0	0	1	0	1	0	0	0	0	0
	0	0	1	0	1	1	0	0	0	0
	0	0	1	1	0	0	0	0	0	1
	0	0	1	1	0	1	0	0	0	0
	0	0	1	1	1	0	0	0	0	0
	0	0	1	1	1	1	0	0	0	0
• • •										
State 101	1	0	1	0	0	0	1	1	0	1
	1	0	1	0	0	1	1	1	0	1
	1	0	1	0	1	0	1	0	0	0
	1	0	1	0	1	1	1	0	0	0
	1	0	1	1	0	0	1	0	0	0
	1	0	1	1	0	1	1	0	0	0
	1	0	1	1	1	0	1	0	0	0
	1	0	1	1	1	1	1	0	0	0

Figure 4-7
Partial Programming Table for Lock.

Formally, the three microinstructions are defined in a source file this way:

```
TITLE Electronic Lock, Definitions.
        WORD  8
CLEAR:  DEF  4X,B#0,3B#000
STEP:   DEF  4X,B#0,3V%:
OPEN:   DEF  4X,B#1,3V%:

        END
```

The TITLE directive simply identifies the file on any listings. The WORD directive defines the width of our microword. In this case our microword is only four bits. Since a single eight bit wide PROM is being used, however, the microword is defined to be eight bits wide.

Notice that each microinstruction is first given a name: "CLEAR:" for example. The colon is not part of the name proper, it is simply a delimiter for identifying the name. The directive DEF indicates that we are to define a microinstruction. The 4X specifies that we do not care about the first four bits of the PROM. When programmed, these bits will be set to a default value. Next, we define a binary constant of one bit wide. This is the output bit, so it is a 0 for CLEAR and STEP, and a 1 for OPEN.

In the case of CLEAR, we will always be returning to state 0. For this reason, we define the last three bits (corresponding to the next state) as constants of value 000. For STEP and OPEN, however, we will want to specify the next state during the programming. We thus specify these three bits as variables. The "%:" symbols are operators that simply tell the meta-assembler that we will be substituting right-justified address bits here. Most microcode assemblers will have a wide variety of such bit-manipulation operators to choose from.

Now that we have defined the microinstructions, let us look at how to actually generate the microcode for our application. Since there are three combinatorial inputs, we know that each state will consist of $2^3 = 8$ microinstructions. We will need to use this value shortly, so we define it as a symbolic constant:

```
SIZE EQU  B#100
```

Our first state is defined and the instructions arranged into a table, just as in the counter example. Here, however, we will use microinstructions in a source file:

```
STATE0:

CLEAR             ; 000 Conditional input.
CLEAR             ; 001 Conditional input.
STEP STATE1/SIZE  ; 010 Conditional input.
DUP 5             ; All other inputs.
CLEAR
```

In this case, the first microinstruction simply puts a 0000 0000 into the first byte of the PROM. The second microinstruction also puts a 0000 0000 into the second byte of the PROM. The third microinstruction is more interesting. The location of this microinstruction corresponds to a conditional input of 010, or in other words button B is being pushed.

Since this is the first sequence in the combination, we want to STEP to STATE1. Notice that we have scaled STATE1 by the value SIZE. This scaling is necessary to accommodate the fact that the next state bits are fed back to the PROM via the high-order address bits. This can be better understood by comparing the lock schematic shown in Figure 4-5 with the PROM organization table shown in Figure 4-7.

The DUP 5 statement is actually a directive. It tells the meta-assembler to duplicate the next line five times. In this case, five CLEAR microinstructions are generated.

The complete source file is shown in Figure 4-8. As a source file, Figure 4-8 is the input to our microcode assembler. Like most assemblers, a microcode assembler typically produces two outputs: an object file and a listing file.

The object file simply contains the bit patterns to be programmed into the PROMs. The listing file is shown in Figure 4-9. Notice that the listing contains more information than the source file: line numbers, addresses, and the object code generated are all shown.

As can be seen from Figure 4-8 and Figure 4-9, microcode listings are anything but easy to follow. The symbolic names for instructions and state address, coupled with the comments, do make the program easier to follow. But it is still an arcane affair. When encountering an unfamiliar microcode listing for the first time, it is often quite useful to prepare a diagram similar to the one in Figure 4-7. This, and remembering that each microinstruction is an address in a PROM, will help one to successfully follow what the source code is trying to convey.

To conclude the discussion of the classic state machine implementation, we will provide some general comments on using PROMs in state machines. As noted in the introduction, we first discussed PROM-based state machines because they are the easiest to follow intuitively. Once one masters the basic concept of the state machine architecture, it becomes clear that the combinatorial block need not be a PROM. A PAL or PLA would work just as well functionally.

It is not as obvious how one would go about synthesizing the equation to produce the correct transfer function. We will look at some of these issues in Chapter 5.

PAL- or PLA-based state machines are often more efficient than the classic PROM-based design in terms of product term utilization and overall speed of operation. The PROM-based designs are still quite useful, however. This is particularly true when very large and complex state machines are needed. Figure 4-10, for example, shows such a case.

The PROM U1 is used to store the next state information. PROMs U3 and U4 store the output data pattern. The effective address for all of the PROMs is generated by the combination of the current state and the

```
        TITLE Electronic Lock, Source File

        ORG    0000         ; Start at 0000.

SIZE:   EQU    B#1000        ; Scale Factor

STATE0:
        CLEAR               ; 000 Conditional  Input.
        CLEAR               ; 001 Conditional  Input.
        STEP   STATE1/SIZE  ; 010 Go to "2".
        DUP 5              ; All other inputs.
        CLEAR

STATE1:
        STEP   STATE2/SIZE  ; 000 Go to "2" (button up).
        CLEAR               ; 001 Conditional  Input.
        STEP   STATE1/SIZE  ; 010 Loop while button is down.
        DUP 5              ; All other inputs.
        CLEAR

STATE2:
        STEP   STATE2/SIZE  ; 000 Wait for button push.
        CLEAR               ; 001 Conditional  Input.
        CLEAR               ; 010 Conditional  Input.
        CLEAR               ; 011 Conditional  Input.
        STEP   STATE3/SIZE  ; 100 Go to "3".
        DUP 3              ; All other inputs.
        CLEAR

STATE3: STEP  STATE4/SIZE    ; 000 Go to "1".
        CLEAR               ;
        CLEAR               ;
        CLEAR               ;
        STEP  STATE3/SIZE    ; 100 Loop while button down.
        DUP 3
        CLEAR

STATE4: STEP   STATE4/SIZE   ; 000 Wait for button push.
        STEP   STATE5/SIZE  ; 001 Go to "1".
        DUP 6              ; All other inputs.
        CLEAR

STATE5:
        OPEN   STATE5/SIZE  ; 000 Stay open while no button is pushed.
        OPEN   STATE5/SIZE  ; 001 Loop while button is down.
        DUP 6              ; All other inputs.
        CLEAR

        END
```

Figure 4-8
Source File for the Electronic Combination Lock.

```
LINE   ADDR   Electronic Lock, Source File
   1                TITLE Electronic Lock, Source File
   2
   3 00000            ORG    0000          ; Start at 0000.
   4
   5          SIZE:  EQU    B#1000         ; Scale Factor.
   6
   7          STATE0:
   8 00000            CLEAR                ; 000 Conditional  Input.
          XXXX0000
   9 00001            CLEAR                ; 001 Conditional  Input.
          XXXX0000
  10 00002            STEP   STATE1/SIZE   ; 010 Go to "2".
          XXXX0001
  11                DUP 5                ; All other inputs.
  12 00003  +         CLEAR
          XXXX0000
  12 00004  +         CLEAR
          XXXX0000
  12 00005  +         CLEAR
          XXXX0000
  12 00006  +         CLEAR
          XXXX0000
  12 00007  +         CLEAR
          XXXX0000
  13
  14          STATE1:
  15 00008            STEP   STATE2/SIZE   ; 000 Go to "2" (button up).
          XXXX0010
  16 00009            CLEAR                ; 001 Conditional  Input.
          XXXX0000
  17 0000A            STEP   STATE1/SIZE   ; 010 Loop while button is down.
          XXXX0001
  18                DUP 5                ; All other inputs.
  19 0000B  +         CLEAR
          XXXX0000
  19 0000C  +         CLEAR
          XXXX0000
  19 0000D  +         CLEAR
          XXXX0000
  19 0000E  +         CLEAR
          XXXX0000
  19 0000F  +         CLEAR
          XXXX0000
  20
  21          STATE2:
  22 00010            STEP   STATE2/SIZE   ; 000 Wait for button push.
          XXXX0010
  23 00011            CLEAR                ; 001 Conditional  Input.
          XXXX0000
  24 00012            CLEAR                ; 010 Conditional  Input.
          XXXX0000
  25 00013            CLEAR                ; 011 Conditional  Input.
          XXXX0000
```

Figure 4-9
Listing File for the Electronic Combination Lock.
(*Figure continues.*)

```
26 00014          STEP   STATE3/SIZE  ; 100 Go to "3".
        XXXX0011
27               DUP 3               ; All other inputs.
28 00015  +        CLEAR
        XXXX0000
28 00016  +        CLEAR
        XXXX0000
28 00017  +        CLEAR
        XXXX0000
29
30 00018   STATE3: STEP  STATE4/SIZE   ; 000 Go to "1".
        XXXX0100
31 00019          CLEAR              ;
        XXXX0000
32 0001A          CLEAR              ;
        XXXX0000
33 0001B          CLEAR              ;
        XXXX0000
34 0001C          STEP  STATE3/SIZE   ; 100 Loop while button down.
        XXXX0011
35               DUP 3
36 0001D  +        CLEAR
        XXXX0000
36 0001E  +        CLEAR
        XXXX0000
36 0001F  +        CLEAR
        XXXX0000
37
38 00020   STATE4: STEP   STATE4/SIZE  ; 000 Wait for button push.
        XXXX0100
39 00021          STEP   STATE5/SIZE  ; 001 Go to "1".
        XXXX0101
40               DUP 6               ; All other inputs.
41 00022  +        CLEAR
        XXXX0000
41 00023  +        CLEAR
        XXXX0000
41 00024  +        CLEAR
        XXXX0000
41 00025  +        CLEAR
        XXXX0000
41 00026  +        CLEAR
        XXXX0000
41 00027  +        CLEAR
        XXXX0000
42
43        STATE5:
44 00028          OPEN   STATE5/SIZE  ; 000 Stay open while no button  is pushed.
        XXXX1101
45 00029          OPEN   STATE5/SIZE  ; 001 Loop while button is down.
        XXXX1101
```

Figure 4-9
(Continued)

```
46              DUP 6           ; All other inputs.
47 0002A  +         CLEAR
          XXXX0000
47 0002B  +         CLEAR
          XXXX0000
47 0002C  +         CLEAR
          XXXX0000
47 0002D  +         CLEAR
          XXXX0000
47 0002E  +         CLEAR
          XXXX0000
47 0002F  +         CLEAR
          XXXX0000
48
49              END


TOTAL ASSEMBLY ERRORS =    0
```

Figure 4-9
(Continued)

conditional inputs. This architecture can easily be expanded to accommodate more states, more conditional inputs, or more outputs. This sort of thing lends itself well to the classic PROM-based architecture.

In the examples presented in this chapter, the PROM has been shown separate from the pipeline register. Again, we have done this for purposes of clarity. In many applications, however, one would choose a PROM that comes with a pipeline register integrated into the device. Such devices are available in a broad range of depths and speeds from AMD, Cypress, and others.

4.3 Chapter Summary

- Programmable state machines are a very regular way of efficiently realizing sequential logic.
- There are two basic types of state machines: the Mealy and Moore.
- A Mealy state machine's outputs respond to changes on the conditional inputs immediately.
- A Moore state machine's outputs change only on clock transitions.
- When programming classic PROM-based state machines, techniques and tools are generally borrowed from the microcoded CPU world.
- A microinstruction can be defined as a set of control bits and the address of the next microinstruction in the microprogram.
- A microprogram is a collection of microinstructions.

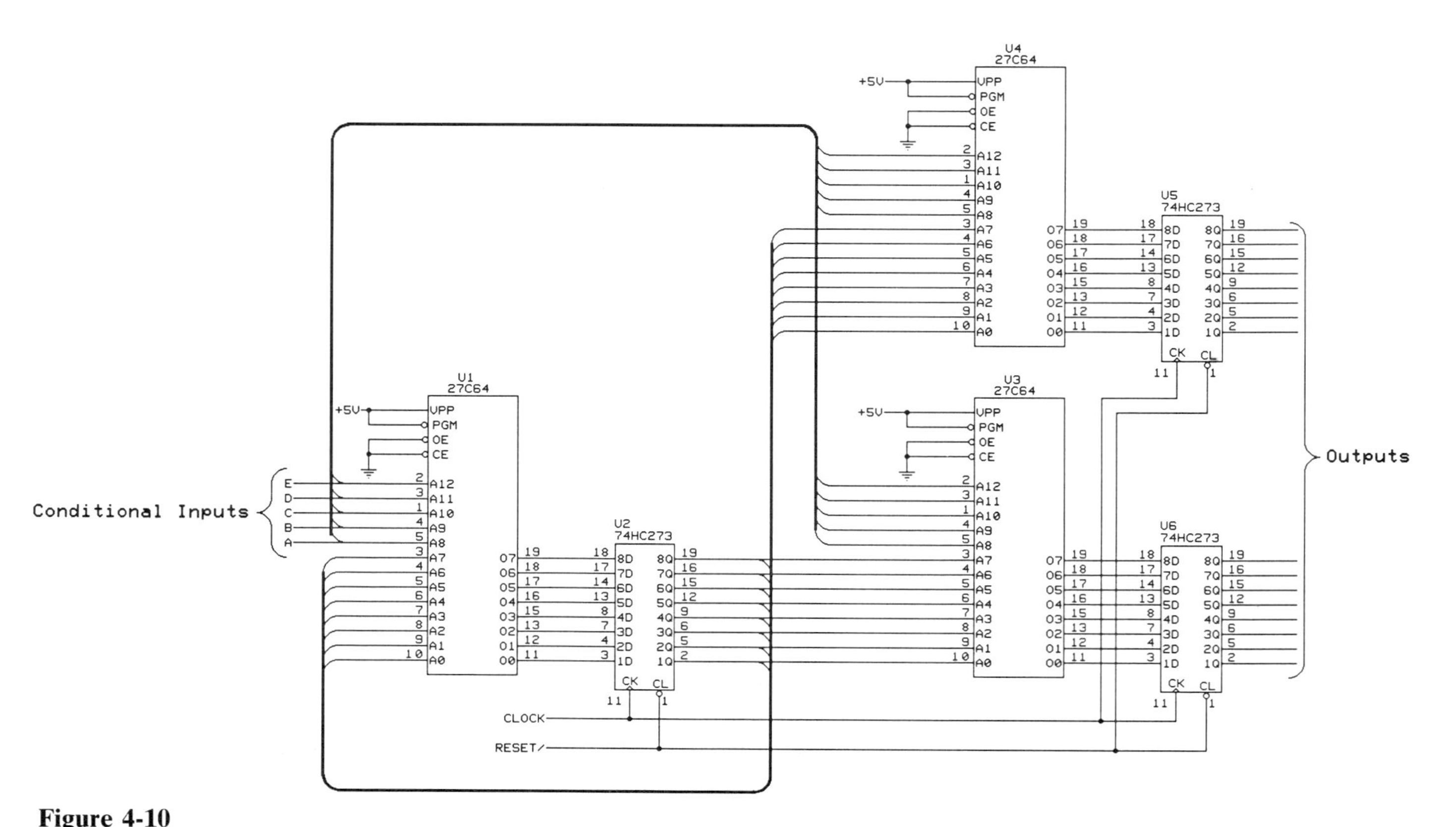

Figure 4-10
A 5 Input, 16 Output, 256 States State Machine.

- The size of a state (the number of addresses required to code the state) is simply 2^n, where n is the number of conditional inputs.
- The number of bits required to encode all the states is: $\mathrm{int}[\log_2(k) + 1]$, where k is the number of states. For example, 6 states will require 3 bits to encode, 14 states will require 4 bits to encode, etc.
- The PROM must be at least 2^{k+n} deep. For example, 4 state bits and 3 conditional inputs will require a PROM of at least 128 words.
- A classic state machine is composed of a PROM and an edge-triggered register set. This is the most flexible state machine implementation but generally not the most efficient (in terms of real estate) implementation.

5

Software Development

In Chapter 4 we briefly introduced the subject of microcoding. In this chapter we will expand the range of the discussion to include some of the other software packages designed for programming PLDs.

As we progress to describing more sophisticated devices, it becomes increasingly difficult to meaningfully discuss the architectural features without also discussing the associated programming. This is true for two reasons. First, without giving actual programming examples, it is difficult to show meaningful applications. Second, for some specialized devices, making an informed choice requires as much understanding of the supporting software as it does of the device itself.

This chapter is divided into three sections. The first is a general discussion of the software associated with PLDs, state machines, and, to a limited extent, microcontrollers. The next section is a light review of some of the general software packages available for working with programmable logic. In the final portion of this chapter we introduce the proLogic compiler by Inlab, Inc.

The proLogic compiler was chosen to provide a uniform platform for the majority of the PLD programming examples. We will look at the reasons we have settled specifically on proLogic after we lay some general groundwork for discussing software packages.

5.1 Introduction to PLD Software

As one might expect, software plays an important role in the field of programmable logic. The software aspect of programmable logic divides into two categories: application software and development software. Application software is the information used to realize the logic function. In other words, the symbolic representation of the function. The role of development software can be defined as converting the symbolic application information (the *source file*) into computer usable binary information (the *object file*).

The goal of application software is to symbolically represent the target logic function in the most efficient way possible. Let us consider a practical example. Assume we have two doors. For whatever reason, we want an alarm to sound when both doors are open at the same time (as in an airlock, for example). In words, our function is "if door A and door B are both open, then sound an alarm." This is the target function.

The function is clear to a human being but in practice there are several problems. First, it contains extraneous information. We do not care, from a logical point of view, what events A and B are. The essence of the function would not change if the doors were actually windows. This extraneous information makes it difficult to efficiently deal with only the logic.

One solution is to represent the function symbolically. For example, we do not lose anything in terms of the logic by simply saying "if A AND B then C" and leaving A, B, and C to be defined. Alternatively, we could simply write:

$$C = A * B$$

The important thing to remember is that both representations are simply symbolic representations of the function.

The point in going through this exercise is to give some insight into the software generation process. Let us look at several ways to realize the function. The first (and simplest) is to use an AND gate as in Figure 5-1a. The second approach is to simply tabulate the desired outputs for all possible inputs. Such a table is illustrated in Figure 5-1b. Finally, we could express the function algorithmicly:

```
Get A
Get B
If A AND B then ALARM
```

The options for building a circuit to actually perform the function directly follow from our choice of how we symbolically represent the function.

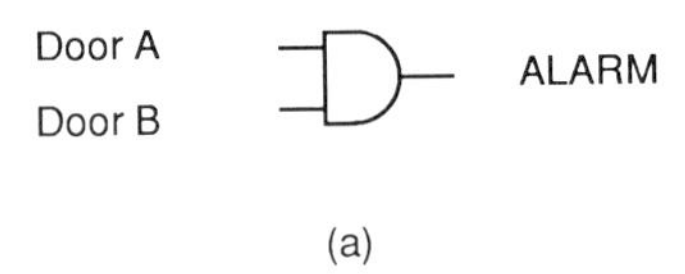

(a)

Door A	Door B	Alarm
0	0	off
0	1	off
1	0	off
1	1	on

(b)

Figure 5-1
A Simple Airlock Alarm

For example, if we represent the function as

$$C = A * B$$

then a natural way to realize the function is with a simple PAL (see Chapter 3). On the other hand, if we tabulate the desired outputs for all possible inputs, we could easily realize the circuit with a purely combinatorial approach by using a simple PROM. Or we could realize the circuit sequentially as a state machine (not that there is any advantage to doing so in this example). If we expressed the function algorithmicly, either a state machine or a microcontroller might be used to realize the circuit.

The main objective of this discussion is to show that the function, its representation, and the engine used to realize the function are all interdependent.

As noted in Chapter 1, the symbolic information in the application software is generally called the source code. It may be a schematic drawing from a CAD file, a timing diagram stored in a file, or, as in the last chapter, an ASCII text file. The output, or binary code, is called the object code.

For any type of PLD the output file is almost always a JEDEC standard file. JEDEC programming files are ASCII text files that tell the PLD programmer exactly what fuse to blow or which cells to program. The file is human readable though it is generally of little interest.

All PLD development software is similar in that it takes the source file and converts it to the JEDEC output file. Where each of the various packages differ is in the form that the data is accepted, the devices covered, and special features. Special features include such things as graphical interfaces, error checking, simulation options, and other specialized functions.

Originally it was easy to partition packages into a simple matrix. The package was either an assembler or a compiler, and it was either provided by the manufacturer or it was a third-party product. We will define these distinctions, but we do so with the warning that the dividing lines are becoming blurred.

PLD assemblers were the first packages available for working with PLDs. As we saw in Chapter 3, there is a rather direct relationship between a logic equation and the fuse pattern used to realize the equation. As long as the equation is in sufficiently simple form, it is a matter of programming the correct cell for either a 1 or a 0 based on the sense of the variable. And that is pretty much what the first PLD assemblers did: they simply converted the sense of the variable to the sense of the fuse. It was up to the engineer to insure that the information was in a "sufficiently simple form."

As PLD assemblers progressed, they became more flexible. They developed the capability of handling symbolic assignments. For example, code could be written as:

```
definitions section:
A = pin 1
B = pin 2
C = pin 3

equations section:
C = A * B.
```

Once the PLD was programmed, a high on pin 1 and a high on pin 2 would cause pin 3 to go high; anything else would cause pin 3 to go low. It was still up to the engineer to insure that the equation could map into the PLD and was in an efficient form. Things like

$$C = A * (B + C)$$

would generally cause an assembler to abort. Worse, some assemblers would take such an equation, but it was anyone's guess what the results would be.

Modern assemblers are very flexible and efficient. They can take a wide range of equation forms, simplify the equation, reduce it to its minimalist form, and develop an optimal programming map for the target PLD. However, by definition a PLD assembler is limited to essentially a one-to-one mapping between input function and the device architecture.

In developing the logic equations to program into a PLD, the engineer is generally working from some other source of input. Often the engineer is working from a timing diagram or a table of requirements. It is a laborious and error prone process to convert these alternate sources into usable logic equations. PLD compilers are capable of doing this conversion work for the engineer. Later, we will demonstrate how the tables produced in Chapters 3 and 4 can be directly input to a compiler.

There are significant advantages to the compiler's ability to use forms of definition closer to natural input of the engineer. First, as noted, translation errors are reduced. Second, design process time is speeded up. The time to go from specification to a working device is decreased. Equally important or possibly more important is the fact that changes can be handled rapidly. A change in a specification table does not mean hours of rederiving logic equations. Finally, the documentation of the design is improved. Rather than simply looking at rows of equations, one can quickly see the intent of the design.

Originally, assemblers were provided free of charge by the manufacturers of the PLDs. Then third-party developers began investing their own time and money to improve the programming tools. These third-party vendors then sold their compilers to users independently of the PLD manufacturers.

The advantage of the assemblers was, of course, that they were free. For the engineer only occasionally designing with PLDs, there was not a great sacrifice in using these relatively limited tools. Unfortunately, the assemblers were limited to programming the manufacturer's own devices. Thus, a design using a variety of PLDs might require several different development packages. This increased the confusion factor and made documentation even more difficult.

Compilers, on the other hand, were designed to support several manufacturers' devices. This made it easier to choose the optimal PLD in the first place and to uniformly document the design process independent of the device selection. Furthermore, the cost of the compilers was rapidly made back in the time saved during the design process.

These distinctions among software types and packages are still very useful. As noted earlier, however, the demarcation lines are not as clear as they were just a few years ago. Many of the manufacturers' packages distributed today are quite sophisticated. And even though these products often retain the categorization as assemblers, they generally possess sophisticated features such as table entry, equation minimization, and simulation.

The distinctions between third-party and manufacturers' software is also blurring. Manufacturers still only distribute software to cover their own products. However, the software distributed is often a third-party product with only the libraries for the manufacturer's products included. Sometimes this software is sold at a discount to the third-party manufacturer's full price, and it is sometimes distributed to users as an incentive to use the PLD manufacturer's devices.

In the next section we will look at some of the packages available. All of the packages discussed work, and they work well. A decision to choose a particular package is generally based on cost, individual preference, and the coverage for the PLDs of interest.

5.2 Specific Software Packages

5.2.1 PALASM

PALASM was one of the first readily available tools for working with PLDs. Much of the early success enjoyed by MMI's PALs is often credited to the availability of this software support package. Over the years PALASM has been improved and expanded. AMD eventually bought out MMI and has continued to support and distribute PALASM. While not quite up to the standards of third-party software, PALASM is still popular among many engineers. PALASM II can be classed as a very sophisticated assembler with many compiler-like features.

Recently AMD has introduced its PALASM 90 software. This package is more glitzy with such features as color waveform output from the simulator. However, the software is rarely distributed for free. It remains to be seen whether or not PALASM 90 will be up to competing with the sophisticated third-party offerings. PALASM, as of the time of this writing, is limited to supporting combinatorial and registered PAL-type PLDs.

5.2.2 AMAZE

AMAZE is the product of Signetics Corporation. Considering that this package is usually provided free to users, it is quite sophisticated. AMAZE provides for a number of compiler-like features and like PALASM can be classed as a near-compiler assembler. It still retains many of the idiosyncracies of the original assemblers, and locating an error in a design can be tricky.

On the other hand, it does provide a simulation capability that is always important for resolving any ambiguities. AMAZE has been around for a long time and has a large user base. Particularly of interest is its capability of defining common products terms. This capability is a direct result of the fact that AMAZE is designed to program PLA and PLA-based sequencers.

5.2.3 CUPL

The first true compiler generally available for PLD design was CUPL. As such, it has gained a strong following among many designers. Its syntax is relatively straightforward, and CUPL is still readily available.

AMD has begun distributing a limited version of CUPL called AmCUPL. The main limitation over the full-blown version is that AmCUPL (naturally) only supports AMD and MMI parts. CUPL's favor seems to come and go, but its admirers are a loyal lot. CUPL will probably be around for awhile.

5.2.4 ABEL

Data I/O is one of the largest and best known of the PLD programmer manufacturers. Recognizing the value of compilers, Data I/O developed its ABEL compiler. It is always an advantage to have as integrated a package as possible, and Data I/O has capitalized on this by providing both the hardware and software to program PLDs. Like CUPL, ABEL is not an inexpensive item by today's standards. However, it is probably as close to an industry standard PLD compiler as one gets.

ABEL, like Data I/O's programmers, will support most any programmable device. This is particularly handy when working in an environment where many different PLDs will be encountered. ABEL has two major limitations: its high cost and the fact that it comes with a hardware interlock that allows it to run on only one machine at a time. This last point is somewhat sensitive. A program that is licensed for a single sight is not really restricted by having a hardware interlock, since it should only be used on one machine at a time. On the other hand, the *last* thing any development lab needs is another gadget to get in the way, breakdown, or corrupt the hardware.

5.2.5 PLDesigner

With its PLDesigner software, MINC has become known as the source of the "cadillac of PLD software." Several very useful features rarely found in other packages are provided by MINC's offering. Among these are the very sophisticated and flexible graphics interface, the ability to defer parts selection until after the design process, automatic partitioning of the design, and a simple built-in expert system.

With most design packages, the user must first choose which PLD will be used. The package then compiles a fuse list for that particular device. While this is not generally a problem, it is much more logical to choose a device *after* the compilation is done. This allows the engineer to select the device of the minimum complexity that is required to do the job, thus saving money, power, and real estate.

This ability to aid the engineer in selecting the simplest device is probably more important than it first seems. It is not at all uncommon for engineers to get into a comfortable rut using only a few PLD types. This ability for one device to fill many different functions is in fact one the advantages of programmable logic—so we are not knocking it. However, it can lead to large overkill in many designs. When one compiles the design first and then selects a package, this tendency is minimized.

Another significant advantage is that large designs can be compiled and simulated, even if they are too large to fit into a single device. The partitioning is more likely to be seamless if the entire design has been developed and simulated in a single file.

Like CUPL and ABEL, however, PLDesigner is not inexpensive. Prices range from $1,500 to over $5,000 at the time this is being written. Of course, if the tool saves even a week or two of a designer's time, it will rapidly pay for itself. Alas, in these days of low-cost PC software, the price will put off many managers responsible for authorizing software purchases.

5.2.6 Plan II

PLAN II is an interesting package that is distributed free of charge by National Semiconductor. PLAN II supports not only the common PALs but also supports GALs. Other interesting capabilities include the ability to *input* JEDEC files and convert them into logic equations. This may not seem all that important since the flow is exactly the opposite of what an assembler normally does, but it can be quite useful when engineering has lost the source files, production only has the object files, and changes are needed.

PLAN II is simple to use. However, it does not have a simulator. This limitation is significant and precludes the use of PLAN II for most serious applications.

5.2.7 Device Specific Packages

As noted earlier, the software available for a device may be more important than the device itself. Device specific software is generated for a variety of reasons. In some cases, it is simply that the device manufacturer feels only a unique program is capable of making the best use of the product. This is true of the advanced PLDs from Xilinx and Altera discussed in Chapter 6.

Some devices, such as the synchronous address multiplexer (SAM) from Wafer Scale Integration, have such a unique architecture that only a unique software package can generate the programming necessary for such a part.

As PLDs become more sophisticated, the need for specialized development packages increases. At the same time, there is a general trend in the industry to open up previously specialized devices to the programming tools of third-party software developers.

These factors must all be taken into account when selecting a programmable device for a particular job. One must factor in not only the cost of the device but the cost of obtaining whatever software is necessary to easily and efficiently program it, as well as the cost of learning both the device and the software.

No matter how sophisticated or powerful a PLD, it is just so much fused sand if it cannot be efficiently programmed to accomplish the task at hand.

5.3 The proLogic Compiler

Of all of the packages available, we have standardized on Inlab's proLogic for our examples in this text. There are a variety of reasons for this:

- Inlab's proLogic is a third-party product. As such, it covers a variety of manufacturer's devices.
- It is a very simple package to understand and use. This simplicity frees us from devoting an inordinate amount of time to discussing the specifics of the software.
- For all of its simplicity, proLogic is a powerful and sophisticated tool. Its compiler-level features free us from many of the idiosyncracies plaguing less sophisticated packages.
- proLogic makes use of a C-like syntax. Thus its syntax will be familiar to a wide audience. proLogic programs are for the most part readable without needing to understand the proLogic syntax.
- While proLogic is not a "bells and whistles" type of program, it does have all of the tools and features one needs in a compiler. What the reader learns in studying the examples in this text will for the most part be transportable to other PLD languages.

- proLogic is inexpensive ($250.00 at press time). Furthermore, a version with a TI-only library is distributed by TI. Generally, the TI package is available free of charge.

In this section we will discuss how the proLogic compiler is used in several examples. The idea is to show the key concepts involved in programming a PLD; we are not specifically trying to document the syntax and use of proLogic. For those interested in using proLogic, the mechanics of file handling and the nuances of use are documented in the compiler's user manual. For those who will be using other software packages, most of what we present here will be directly applicable. It will simply be a matter of adapting the general techniques mentioned in this text to the specifics of the development software being used.

One point should be emphasized, however: proLogic is a friendly and forgiving tool. This is not as true of many of the other software development packages. Often, subtle syntax issues can make or break a PLD design.

As noted previously, all development software has one simple objective: to convert symbolic logic information into the JEDEC file necessary to actually program a PLD.

A variety of steps are required to accomplish this transformation:

- We must find some way of relating the symbolic information of the design to the physical location of the PLDs resources. These resources can be pins, registers, special functions, and other features of the device.
- We need to specify exactly what we want the device to do. This specification may take a variety of forms depending upon the sophistication of the development tool. Generally, the form will be logic equations, but proLogic also supports logic tables and state tables.
- The design specification generally will not be in the minimalist form. Thus, the design needs to be reduced to its simplest form to insure the highest probability that it will fit into the target device. proLogic, as well as other high-quality compilers, will do this automatically.
- This design information must then be converted to a JEDEC file for programming the device.
- It is one thing to specify a design and quite another to insure that what we specified is what we really want. Simulation is the vehicle to prove that what we told the compiler we wanted is *actually* what we get. Simulation is sometimes handled as an afterthought to the design process. This can be a serious mistake. Simulation is one of the most important tools in any development software suite.

5.3.1 Pin Names and Signal Conventions

Many of the programming languages make use of a table at the start of the source file to specify the symbolic names of the pins in the PLD. proLogic takes a more direct approach to the problem. Key nodes in the PLD are assigned signal names. These names are fixed by the software but are generally intuitively obvious. For example, pin number one is simply "pin1."

Various operators are available for working with the signal names. These are similar to the operators found in C:

! is the complement operator

& is the AND operator

| is the OR operator

Thus if one chooses to implement a simple two input NAND gate with a PAL16L8, one may simply write:

```
!pin19 = pin2 & pin3;
```

Like C, the semicolon at the end of the line tells the compiler that the logical end of line has been reached. The semicolon is mandatory.

For simple designs there is really only one other thing we need to know. That is how to specify what device we want to use. We do this with the "include" directive.
To compile our file, only two lines are required:

```
include P16L8;
!pin19 = pin2 & pin3;
```

This simple two line file will compile, yielding a two input NAND gate. This is about as simple and straightforward as programming a PLD gets.

Working directly with signal names directly has its advantages. However, as we demonstrated in the beginning of the chapter, it is often desirable to have names with more meaning. We can accomplish this by defining a symbol that translates into the signal name. For example:

```
include P16L8;

define ADDR0 = pin2;
define ADDR1 = pin3;
define ENABLE = !pin19;

ENABLE = ADDR0 & ADDR1;
```

This will produce exactly the same output as our previous file. It is much easier to read, however. If someone needs to modify the file later, it will be much easier to see what we were trying to accomplish. The readability of the file can be enhanced even further if we add comments. As in C, this can be done by enclosing our comments in the comment operators:

```
/* This is valid comment line. */
```

5.3.2 Truth Tables

Truth tables are a very useful way of specifying many logic functions. When the information is presented in tabular form, the purpose and function may be far clearer than simply looking at sets of logic equations. For example, in Chapter 3 we showed how the seven-segment decoder could be implemented with a PROM. If we wanted to implement this device with a PAL16L8 for example, we could simply code this directly as shown in Figure 5-2.

Notice the form of the table. The left-hand side is simply the input variable. The right-hand side contains the outputs. The two are divided by the ":" operator so that the compiler knows which is which. Underneath the table headings are the associated values for each input and output. When we compile the listing in Figure 5-2, the compiler automatically generates and simplifies the equations necessary for the design. Figure 5-3 shows the equations proLogic generated from this seven-segment decoder table. We have left out such features as blanking and display test to keep the example simple. However, these features could have been added by simply expanding the table.

The compiler also produces other outputs. One of these is the fuse map shown in Figure 5-4. Fuse maps are of much less direct interest, but we have included this one to show what they look like. Notice that the fuse map provides an immediate visual feel for the percentage of the PAL16L8's resources that have been used.

Fuse maps will occasionally be useful for debugging complex designs. If one cannot spot what is going wrong in the source file, looking at the equations list and the fuse map can sometimes provide valuable hints pointing to the source of the problem.

5.3.3 State Diagrams

Another extremely useful feature of the proLogic compiler is the ability to handle state diagrams. State diagrams have the following general syntax:

```
state_diagram STATE_VARIABLES {
     state = VALUE TRANSITION_TERM
                                     }
```

```
/******************************************************************************/
/*            7 Segment LED Decoder.                                          */
/*                                                                            */
/* Device: PAL 16L8                                                           */
/* Function: Decodes a binary input for display on a 7 segment LED.   */
/*                                                                            */
/* Created: July 1st, 1990                                                    */
/* Engineer: Jim Broesch.                                                     */
/*                                                                            */
/* This file is coded for compilation by proLogic.                           */
/******************************************************************************/
include P16L8;

/* Input Definitions. */

define BIT3 = pin1;
define BIT2 = pin2;
define BIT1 = pin3;
define BIT0 = pin4;

/* Output definitions. */

define SEGG = !pin19;
define SEGF = !pin18;
define SEGE = !pin17;
define SEGD = !pin16;
define SEGC = !pin15;
define SEGB = !pin14;
define SEGA = !pin13;

/* The truth table for the decoding follows. A 1 in the segment      */
/* column indicates that the segment will be on. Since the           */
/* outputs of the 16L8 are active low, the pin will be a zero if there */
/* is a 1 in the column                                               */

truth_table {
      BIT3 BIT2 BIT1 BIT0 : SEGG SEGF SEGE SEGD SEGC SEGB SEGA ;
/* 0 */  0  0  0  0 : 0  1  1  1  1  1  1 ;
/* 1 */  0  0  0  1 : 0  0  0  1  1  0  0 ;
/* 2 */  0  0  1  0 : 1  1  1  0  1  1  0 ;
/* 3 */  0  0  1  1 : 1  0  1  1  1  1  0 ;
/* 4 */  0  1  0  0 : 1  0  0  1  1  0  1 ;
/* 5 */  0  1  0  1 : 1  0  1  1  0  1  1 ;
/* 6 */  0  1  1  0 : 1  1  1  1  0  1  1 ;
/* 7 */  0  1  1  1 : 0  0  0  1  1  1  0 ;
/* 8 */  1  0  0  0 : 1  1  1  1  1  1  1 ;
/* 9 */  1  0  0  1 : 1  0  0  1  1  1  1 ;
/* A */  1  0  1  0 : 0  1  1  1  1  1  1 ;
/* B */  1  0  1  1 : 1  1  1  1  0  0  1 ;
/* C */  1  1  0  0 : 0  0  1  1  1  1  0 ;
/* D */  1  1  0  1 : 1  1  1  1  1  0  0 ;
/* E */  1  1  1  0 : 1  1  1  0  0  1  1 ;
/* F */  1  1  1  1 : 1  1  0  0  0  1  1 ;
   }

/* Notice that the sense of the pin is inverted with respect   */
/* to the columns above. This is due to the active low outputs     */
/* of the 16L8.                                                 */

test_vectors {
      pin1 pin2 pin3 pin4 pin19 pin18 pin17 pin16 pin15 pin14 pin13 ;
       0    0    0    0    H     L     L     L     L     L     L   ;
       0    0    0    1    H     H     H     L     L     H     H   ;
       0    0    1    0    L     L     L     H     L     L     H   ;
       0    0    1    1    L     H     L     L     L     L     H   ;
       0    1    0    0    L     H     H     L     L     H     L   ;
       0    1    0    1    L     H     L     L     H     L     L   ;
       0    1    1    0    L     L     L     L     H     L     L   ;
       0    1    1    1    H     H     H     L     L     L     H   ;
       1    0    0    0    L     L     L     L     L     L     L   ;
       1    0    0    1    L     H     H     L     L     L     L   ;
       1    0    1    0    H     L     L     L     L     L     L   ;
       1    0    1    1    L     L     L     L     H     H     L   ;
       1    1    0    0    H     H     L     L     L     L     H   ;
       1    1    0    1    L     L     L     L     L     H     H   ;
       1    1    1    0    L     L     L     H     H     L     L   ;
       1    1    1    1    L     L     H     H     H     L     L   ;
   }
```

Figure 5-2

LED Decoder Source Code.

```
proLogic Compiler
Texas Instruments V1.97
Copyright (C) 1989 INLAB, Inc.
Signal Specifications

pin18.oe= 1

pin17.oe= 1

pin16.oe= 1

pin15.oe= 1

pin14.oe= 1

pin13.oe= 1

pin19.oe= 1

!pin13= !pin1 & pin2 & !pin3 | !pin2 & !pin3 & !pin4
        | pin2 & pin3 & !pin4 | pin1 & pin3 | pin1 & !pin2

!pin14= !pin1 & pin2 & pin4 | pin1 & !pin2 & !pin3
        | pin1 & !pin4 | pin2 & pin3 | !pin1 & pin3 | !pin2 & !pin4

!pin15= !pin1 & pin3 & pin4 | pin1 & !pin3
        | !pin3 & !pin4 | !pin2 & !pin4 | !pin1 & !pin2

!pin16= pin1 & !pin2 | !pin1 & pin2 | !pin1 & pin4 | !pin3

!pin17= pin2 & !pin3 & pin4 | pin1 & !pin4
        | pin3 & !pin4 | !pin2 & pin3 | !pin2 & !pin4

!pin18= pin1 & pin2 & pin4 | pin1 & pin3 | pin3 & !pin4
        | !pin2 & !pin4

!pin19= !pin1 & !pin2 & pin3 | !pin1 & pin2 & !pin3
        | pin1 & !pin2 & !pin3 | pin2 & pin3 & !pin4 | pin1 & pin4
```

Figure 5-3
Logic Equations Generated by the proLogic Compiler.

The state variables are registers with internal feedback. State has the same meaning attached to it as in Chapter 4, but it should be noted that this is basically a symbolic name. The value of the state is assigned by VALUE. TRANSITION_TERM specifies the next state. This is most easily demonstrated with a simple example. Consider a simple two bit counter:

```
state_diagram BIT1, BIT0 {
     state S0 = 00 S1;
     state S1 = 01 S2;
     state S2 = 10 S3;
     state S3 = 11 S0;
                        }
```

This code will form a synchronous two bit counter that counts 00 01 10 11 and then cycles back to 00 etc.

State diagrams save considerable time and effort in the design process. While it is not a major problem to derive the equations for the example above, it *is* a tedious and error prone process. Worse, any

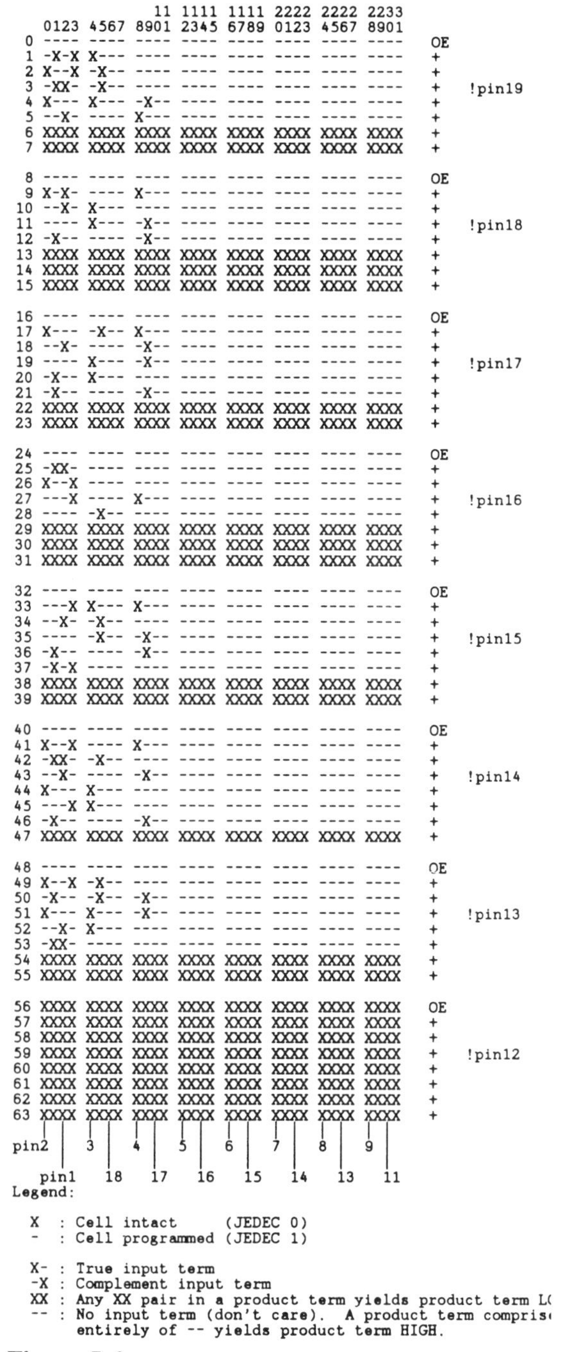

```
                   11 1111 1111 2222 2222 2233
    0123 4567 8901 2345 6789 0123 4567 8901
 0 ---- ---- ---- ---- ---- ---- ---- ----   OE
 1 -X-X X--- ---- ---- ---- ---- ---- ----   +
 2 X--X -X-- ---- ---- ---- ---- ---- ----   +
 3 -XX- -X-- ---- ---- ---- ---- ---- ----   +    !pin19
 4 X--- X--- -X-- ---- ---- ---- ---- ----   +
 5 --X- ---- X--- ---- ---- ---- ---- ----   +
 6 XXXX XXXX XXXX XXXX XXXX XXXX XXXX XXXX   +
 7 XXXX XXXX XXXX XXXX XXXX XXXX XXXX XXXX   +

 8 ---- ---- ---- ---- ---- ---- ---- ----   OE
 9 X-X- ---- X--- ---- ---- ---- ---- ----   +
10 --X- X--- ---- ---- ---- ---- ---- ----   +
11 ---- X--- -X-- ---- ---- ---- ---- ----   +    !pin18
12 -X-- ---- -X-- ---- ---- ---- ---- ----   +
13 XXXX XXXX XXXX XXXX XXXX XXXX XXXX XXXX   +
14 XXXX XXXX XXXX XXXX XXXX XXXX XXXX XXXX   +
15 XXXX XXXX XXXX XXXX XXXX XXXX XXXX XXXX   +

16 ---- ---- ---- ---- ---- ---- ---- ----   OE
17 X--- -X-- X--- ---- ---- ---- ---- ----   +
18 --X- ---- -X-- ---- ---- ---- ---- ----   +
19 ---- X--- -X-- ---- ---- ---- ---- ----   +    !pin17
20 -X-- X--- ---- ---- ---- ---- ---- ----   +
21 -X-- ---- -X-- ---- ---- ---- ---- ----   +
22 XXXX XXXX XXXX XXXX XXXX XXXX XXXX XXXX   +
23 XXXX XXXX XXXX XXXX XXXX XXXX XXXX XXXX   +

24 ---- ---- ---- ---- ---- ---- ---- ----   OE
25 -XX- ---- ---- ---- ---- ---- ---- ----   +
26 X--X ---- ---- ---- ---- ---- ---- ----   +
27 ---X ---- X--- ---- ---- ---- ---- ----   +    !pin16
28 ---- -X-- ---- ---- ---- ---- ---- ----   +
29 XXXX XXXX XXXX XXXX XXXX XXXX XXXX XXXX   +
30 XXXX XXXX XXXX XXXX XXXX XXXX XXXX XXXX   +
31 XXXX XXXX XXXX XXXX XXXX XXXX XXXX XXXX   +

32 ---- ---- ---- ---- ---- ---- ---- ----   OE
33 ---X X--- X--- ---- ---- ---- ---- ----   +
34 --X- -X-- ---- ---- ---- ---- ---- ----   +
35 ---- -X-- -X-- ---- ---- ---- ---- ----   +    !pin15
36 -X-- ---- -X-- ---- ---- ---- ---- ----   +
37 -X-X ---- ---- ---- ---- ---- ---- ----   +
38 XXXX XXXX XXXX XXXX XXXX XXXX XXXX XXXX   +
39 XXXX XXXX XXXX XXXX XXXX XXXX XXXX XXXX   +

40 ---- ---- ---- ---- ---- ---- ---- ----   OE
41 X--X ---- X--- ---- ---- ---- ---- ----   +
42 -XX- -X-- ---- ---- ---- ---- ---- ----   +
43 --X- ---- -X-- ---- ---- ---- ---- ----   +    !pin14
44 X--- X--- ---- ---- ---- ---- ---- ----   +
45 ---X X--- ---- ---- ---- ---- ---- ----   +
46 -X-- ---- -X-- ---- ---- ---- ---- ----   +
47 XXXX XXXX XXXX XXXX XXXX XXXX XXXX XXXX   +

48 ---- ---- ---- ---- ---- ---- ---- ----   OE
49 X--X -X-- ---- ---- ---- ---- ---- ----   +
50 -X-- -X-- -X-- ---- ---- ---- ---- ----   +
51 X--- X--- -X-- ---- ---- ---- ---- ----   +    !pin13
52 --X- X--- ---- ---- ---- ---- ---- ----   +
53 -XX- ---- ---- ---- ---- ---- ---- ----   +
54 XXXX XXXX XXXX XXXX XXXX XXXX XXXX XXXX   +
55 XXXX XXXX XXXX XXXX XXXX XXXX XXXX XXXX   +

56 XXXX XXXX XXXX XXXX XXXX XXXX XXXX XXXX   OE
57 XXXX XXXX XXXX XXXX XXXX XXXX XXXX XXXX   +
58 XXXX XXXX XXXX XXXX XXXX XXXX XXXX XXXX   +
59 XXXX XXXX XXXX XXXX XXXX XXXX XXXX XXXX   +    !pin12
60 XXXX XXXX XXXX XXXX XXXX XXXX XXXX XXXX   +
61 XXXX XXXX XXXX XXXX XXXX XXXX XXXX XXXX   +
62 XXXX XXXX XXXX XXXX XXXX XXXX XXXX XXXX   +
63 XXXX XXXX XXXX XXXX XXXX XXXX XXXX XXXX   +
pin2    3    4    5    6    7    8    9
   pin1   18   17   16   15   14   13   11
Legend:

  X  : Cell intact      (JEDEC 0)
  -  : Cell programmed  (JEDEC 1)

  X- : True input term
  -X : Complement input term
  XX : Any XX pair in a product term yields product term L(
  -- : No input term (don't care).  A product term compris(
       entirely of -- yields product term HIGH.
```

Figure 5-4
proLogic Fuse Map for the Seven-Segment Decoder.

change in the state diagram means that the whole derivation process must be executed again from the beginning.

Conditional statements can be added to the basic state diagram. For example, if we wanted the counter to hold in state zero upon command, we could use a line such as:

```
state S0 = 00 if (!HOLD) S1;
```

If HOLD is true, the conditional IF statement is false, and the counter will not transition to S1. If HOLD is false, the conditional statement is true, and the counter will sequence to S1.

Conditional statements can be placed at the start of the state definition. This is particularly useful for things such as resets:

```
state_diagram BIT1, BIT0 {
      if (RESET) S0;
      state S0 = 00 S1;
      state S1 = 01 S2;
      state S2 = 10 S3;
      state S3 = 11 S0;
                         }
```

This code will cause a synchronous reset of the counter if RESET is true. Or in other words, the next state transition will be to S0, *regardless of the current state,* if RESET is true. Conditional statements can also take the if-else form, just as in C.

5.3.4 PAL Version of the Four Bit Counter

We can put this all together in a practical example. Once again we will revisit our push button counter, this time implemented with a PAL16R4 and a PAL16L8. The PAL16R4 shown in Figure 5-5 is similar to the PAL16L8. However, four latches have been added to the outputs. The four registers of the PAL16R4 make it a natural for the counter. We will not let the combinatorial outputs go to waste either; we will use them to build the debouncing RS latch.

The output of the PAL16R4 will be decoded by the PAL16L8 as in the previous example.

The source file for the PAL16R4 is shown in Figure 5-6. A schematic of the counter is shown in Figure 5-7. Several particularly interesting points can be found in the implementation:

1. As with the state machine example, we have made our counter an up/down implementation.

PAL16R4AM, PAL16R4A-2M
STANDARD HIGH-SPEED *PAL*® CIRCUITS

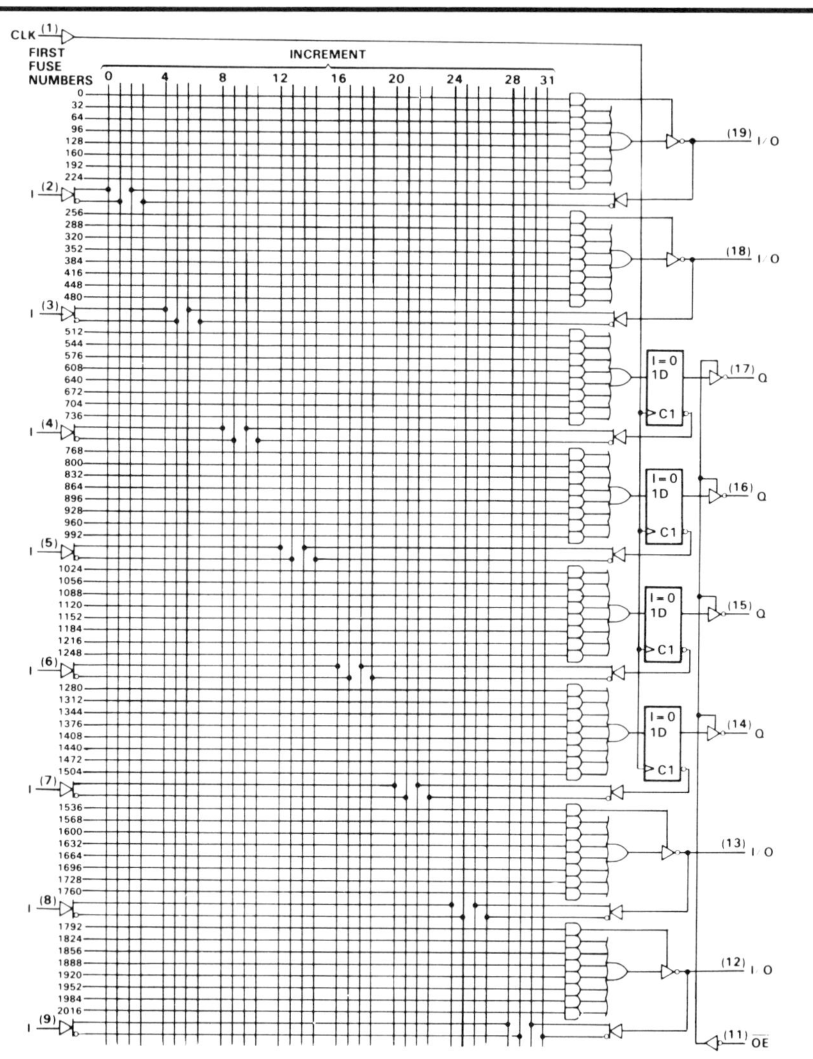

Figure 5-5
The PAL16R4. (Reprinted with permission of Texas Instruments, ©1990.)

2. We are using the on-board combinatorial logic to generate the debounced clock for the register's synchronous clock.
3. The reset is a synchronous one. This means that to reset the counter we must push and hold RESET while pushing the count button. This is not a particularly desirable feature since an asynchronous reset is more natural. It is the best we can do with a 16R4, however.

After we have successfully compiled the source file in Figure 5-6, we are ready for the next step: simulation.

```
/*******************************************************************/
/*                      4 Bit Binary Counter.                      */
/* Device: PAL16R4                                                 */
/* Function: Synchronous 4 bit counter with debouncing             */
/*           latch.                                                */
/* Created: July 5, 1990                                           */
/* Engineer: Jim Broesch                                           */
/*                                                                 */
/* This file is coded for compilation by proLogic.                 */
/*******************************************************************/
include p16r4;

/* Input Definitions. */
define CLOCK = pin1;
define RESET = pin2;
define UP    = pin3;
define OE    = pin11;

/* Output Definitions. */

define D3 = pin17;
define D2 = pin16;
define D1 = pin15;
define D0 = pin14;

define SET   = pin4;
define CLEAR = pin5;

define QOUT = pin19;
define QBAR = pin18;

/* These are the logic equations for the debouncing latch. */

!QOUT = SET & QBAR;
!QBAR = CLEAR & QOUT;

/* Now comes the state table for the counter. */

state_diagram !D3,!D2,!D1,!D0 {
     if (!RESET) s0;
     state s0 = 0000 {if (UP) s1; else sF;}
     state s1 = 0001 {if (UP) s2; else s0;}
     state s2 = 0010 {if (UP) s3; else s1;}
     state s3 = 0011 {if (UP) s4; else s2;}
     state s4 = 0100 {if (UP) s5; else s3;}
     state s5 = 0101 {if (UP) s6; else s4;}
     state s6 = 0110 {if (UP) s7; else s5;}
     state s7 = 0111 {if (UP) s8; else s6;}
     state s8 = 1000 {if (UP) s9; else s7;}
     state s9 = 1001 {if (UP) sA; else s8;}
     state sA = 1010 {if (UP) sB; else s9;}
     state sB = 1011 {if (UP) sC; else sA;}
     state sC = 1100 {if (UP) sD; else sB;}
     state sD = 1101 {if (UP) sE; else sC;}
     state sE = 1110 {if (UP) sF; else sD;}
     state sF = 1111 {if (UP) s0; else sE;}
                                  }
```

Figure 5-6
Four Bit Binary Counter. (*Figure continues.*)

```
/* The following vectors perform an abbreviated functional test */
/*  on the counter.                                              */

test_vectors {   /* Test the up/down counter logic. */
  CLOCK RESET UP OE  D3  D2  D1  D0  ;
    c     1    0  1   Z   Z   Z   Z   ;
    c     0    1  0   L   L   L   L   ;
    c     1    0  0   H   H   H   H   ;
    c     0    0  0   L   L   L   L   ;
    c     1    1  0   L   L   L   H   ;
    c     1    1  0   L   L   H   L   ;
    c     1    1  0   L   L   H   H   ;
    c     1    1  0   L   H   L   L   ;
    c     1    1  0   L   H   L   H   ;
    c     1    1  0   L   H   H   L   ;
    c     1    1  0   L   H   H   H   ;
    c     1    1  0   H   L   L   L   ;
    c     1    1  0   H   L   L   H   ;
    c     1    1  0   H   L   H   L   ;
    c     1    1  0   H   L   H   H   ;
    c     1    1  0   H   H   L   L   ;
    c     1    1  0   H   H   L   H   ;
    c     1    1  0   H   H   H   L   ;
    c     1    1  0   H   H   H   H   ;
    c     1    1  0   L   L   L   L   ;
    c     1    0  0   H   H   H   H   ;
    c     1    0  0   H   H   H   L   ;
    c     1    0  0   H   H   L   H   ;
    c     1    0  0   H   H   L   L   ;
    c     1    0  0   H   L   H   H   ;
    c     1    0  0   H   L   H   L   ;
    c     1    0  0   H   L   L   H   ;
    c     1    0  0   H   L   L   L   ;
    c     1    0  0   L   H   H   H   ;
    c     1    0  0   L   H   H   L   ;
    c     1    0  0   L   H   L   H   ;
    c     1    0  0   L   H   L   L   ;
    c     1    0  0   L   L   H   H   ;
    c     1    0  0   L   L   H   L   ;
    c     1    0  0   L   L   L   H   ;
    c     1    0  0   L   L   L   L   ;
             }

/* The next vectors test the latch. */

test_vectors{   /* Test the RS switch debouncing latch. */
    CLEAR SET QOUT QBAR ;
      0    1   L    H   ;
      1    1   L    H   ;
      1    0   H    L   ;
      1    1   H    L   ;
           }
```

Figure 5-6
(*Continued*)

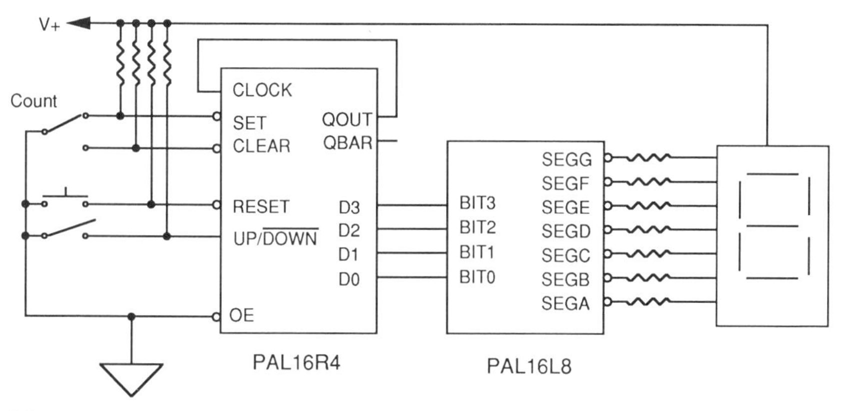

Figure 5-7
PAL Version of the Four Bit Counter.

5.3.5 Simulation

Most PLD assemblers and compilers have some type of simulation capability. Any development packages that do not have simulation capability are to be avoided at all costs! Even for simple applications, the number of obscurities that can confuse the design process are practically unlimited.

For example, the application we have just presented is not particularly aggressive by modern standards. Nevertheless, it took the author several iterations after getting rid of syntax errors to work out the correct coding.

At the end of each iteration the design was simulated. Each time the design failed, the source file was modified, recompiled, and resimulated. When the design finally passed simulation, the actual parts were programmed. When prototyped on a breadboard, the actual circuit was a first-pass success.

Had we simply programmed a part, then tried to figure out what was wrong by plugging the part into the breadboard and observing the circuit, a two hour job would have been more like a two day effort. If we had been using fuse-based PLDs, we would also have used up a fair number of valuable parts.

Simulation is often used to resolve questions of ambiguity in syntax and logic. For example, consider the outputs on a PAL16L8. We know that these outputs are inverted or active low. So what if we have a code like:

```
define A = pin1;
define B = !pin19;

B = A;
```

Will *B* be the complement of *A* or will it be the true form of *A*? The answer is, unfortunately, "it depends." It depends on the assembler or the compiler and sometimes on which revision of the compiler or assembler is being used.

So, how do we go about deciding which is the case for our application? There are two possible courses of action (outside of just asking someone who might have the answer): program the part and try it or simulate the design. The first option is to be avoided; if we have a fuse-based part, it will be wasted if we are wrong. Even if we have a reprogrammable part, such as the ones we will be discussing in Chapter 6, it is still inefficient in terms of time (and therefore money) to bench test each iteration.

The preferable course of action is to simulate the design before we ever put it in the circuit. In simulating a design, the software essentially models the programmed device. The simulator then applies the inputs we specify, called *test vectors,* and tests to see if we get the output that was specified.

The first step in simulation is to define the test vectors. Test vectors are simply the stimulus applied to the input pins during test. During simulation, the first stimulus is applied and the output is checked against the expected output. This process is repeated until all vectors have been executed.

Test vectors are composed of driving conditions. Some of the driving conditions are:

1 for a logical high input.

0 for a logical low input.

X for a "don't care."

C for a low-high-low clock cycle.

K for a high-low-high clock cycle.

Test conditions can be specified as:

L for a low output.

H for a high output.

Z for a output that should be in the high-impedance state.

N do not test the output.

To include test vectors in our compiled file we simply put in a statement like:

```
test_vectors {
        pin1  pin19 ;
         0      H   ;
         1      L   ;
                 }
```

For proLogic, the test_vectors statement only includes the test vectors in the compiled file. It does not actually model the design or apply the test vectors. This is done by a separate simulator.

The simulator will read the fuse data in from the JEDEC output file, simulate the design, and apply the test vectors. If the output matches what we specified, the proLogic simulator reports no errors. If the output does not match our specification, an error message is generated and the failed vectors are flagged.

Now, to answer the question of what we will get for $A = B$. For version 1.97 of proLogic, the answer is that pin 19 will be the complement of pin 1. If pin 1 is high, pin 19 will be low. If pin 1 is a high, pin 19 will be low. This business of determining the actual sense of a signal is quite common in PLD design and is the reason we digressed in Chapter 3 to discuss why active low signals are so common. Rarely has the author worked on a design project where this type of confusion has not caused some problem. The moral is to check, both from a specification point of view and functionally (i.e., simulation), exactly what the logic is *supposed* to do *and* what it is *actually* doing.

The simulation table of a PAL16R4 is shown in the lower portion of the listing in Figure 5-6. Notice that we have divided the test vectors into two groups. The first group simply exercises the RS latch. The second set of test vectors is used to exercise the counter portion of the design.

This ability to separate test vectors is quite handy. It allows us to concentrate on portions of the design that are logically distinct. This makes the simulation process more modular and the results easier to understand.

For the RS latch, we have tested the design just as we would have on the bench. We first clear the latch by bringing the CLEAR signal LOW. The QOUT signal is tested to see that it goes LOW, and QBAR is tested to see that it goes HIGH. Next, the CLEAR input is brought HIGH to see that the latch retains the correct state. The SET input is then brought LOW, and we test to see that QOUT goes HIGH and QBAR goes LOW. Finally, we bring SET to HIGH and again check to see that the latch contains the correct state.

The simulation for the counter progresses in much the same fashion. We have made use of the clock operator "C" to exercise each state in the counter. Notice that we have cleared the counter at several different points and tested the high-impedance output. We have also made sure that the counter counts both up and down and that both underflow and overflow conditions have been checked.

We should point out that this simulation is a thorough one, but it is not an exhaustive or a rigorous one. We do not, for instance, test the RESET function for each and every possible state. Writing test vectors to do this is not hard but it does get quite tedious. To simplify the process, some software packages provide utilities for automatically generating test vectors.

The sophistication of automatic test vector generators can vary considerably. The simplest ones just apply all possible combinations of inputs to the device. This is handy for combinatorial circuits but obviously is of little use for registered designs. More sophisticated packages will produce a set of vectors designed to insure that every possible transition is tested.

These automatic test vector generators are primarily useful for testing devices in the production environment. They are useful for incoming inspection and for fault isolation when testing suspected bad assemblies. They are not particularly useful to the designer since the vectors are not easy to follow, and their sequence does not necessarily "make sense" from an intuitive standpoint. Even when these tools are available, the designer should still produce his or her own set of test vectors to verify the *design* and use the automatic tools for validation, parts checkout, and production testing.

In Chapter 6 we will see more practical examples of programming PLDs. These examples will show a few of the more sophisticated techniques that can be used to simplify the design process and improve the documentation of a design. In the mean time, we will present a few general comments on software as it relates to PLDs.

5.4 Miscellaneous Comments on Software

The introduction of practical programmable devices has had many impacts on electronics engineering. One of the more interesting effects has been to bring the theoretical and practical aspects of logic design closer together.

Prior to the general acceptance of PLDs, there was only a cursory correlation between the theorems taught in school and the actual designing of a logic circuit. The reason for this was, of course, that the designer had only a fixed number and style of logic elements to work with. One could pick and choose from the parts available in the standard logic families, but beyond that there was little flexibility.

The net effect of this was that a good designer could fit the existing components into the design with the least waste. Logically minimalist solutions were of only academic interest; the working engineer was trying to achieve component-area and cost-efficiency.

Naturally, it is still the major goal of the working engineer to achieve as efficient a design as possible. However, today such a design is much more likely to look like a theoretical logic discussion. As we have seen in the examples in this section, one is now likely to spend a large portion of the design process in specifying the logic requirements in tables, state diagrams, etc. It is up to the compilers to do the actual logic synthesis and only along the rigid lines dictated by the target device's architecture.

There are many advantages to all of this: designs are cleaner and more efficient, products are brought to market faster, and the documentation and reliability of these products is improved.

On the other hand, all progress has its drawbacks as well as its rewards. As noted in Chapter 1, using programmable logic is a trade. The trade is almost always a good one, but let us examine the trade from the visual perspective.

Most engineers are, almost by definition, visually oriented. We think mainly in pictures. The process of designing with discrete gates and individual flip-flops fits well with this orientation. Looking at a well laid out schematic makes it easy to follow the flow of the circuit. One can quickly get an image of what is supposed to be happening. This is particularly true with hierarchial schematics that are well done. The flow and function of the product becomes almost intuitively clear at a quick glance.

On the other hand, as we have seen in the previous examples, the PLD process is often more verbal than visual. We are writing in words or mnemonics what we want *done*. Cognitively, this is in sharp contrast to the more natural process of drawing what we want the machine *to be*.

This concern was not lost on the producers of PLD hardware and software. Thus many software development packages strive to maintain a graphical interface that allows schematic entry of the design. Altera's A+Plus and MAX+Plus packages are good examples of dedicated software packages that allow this capability. Altera provides a library of symbols that are functionally identical to the 7400 series of components.

This is more than just a concession to the more experienced engineers reluctant to master the new skills. Many applications for PLDs are based on replacing existing TTL designs. By providing a library of equivalent components, the translation of these existing designs to the more efficient PLD implementation is greatly simplified. The chance of subtle bugs being introduced into the design are also minimized.

Ultimately, the trend toward a more textual orientation to design will tend to predominate. This will occur simply because logic is inherently a mathematical process, and mathematics is more symbolic than pictorial. Equations and tables will predominate.

This does not mean, however, that schematics are going away. The schematic is still the best way of conveying the actual interconnection and layout of the circuit. Modern design tools will continue to provide improved ways of capitalizing on both textual and schematic design techniques.

5.5 Chapter Summary

- Finding a PLD that meets the design requirements from an architectural perspective is only part of the design process. One must also find supporting software that can be used *efficiently* in the development process.

- Development software can generally be obtained from the PLD vendors for little or no cost.
- Third-party software is also available. While it will generally cost more than vendor-supplied software, third-party software is usually more powerful, covers multiple manufacturers' products, and is often a good investment.
- Specialized devices often require specialized development software. These costs, both in dollars and in time (acquisition time, learning curves, etc.) must be factored into the overall program plan.
- PLD development software should have at least three key features: multiple input formats, logic minimization, and simulation capability.

6

Advanced Forms of PLDs

The devices and architectures discussed so far have been relatively simple and straightforward. These "first generation" devices were designed primarily to replace awkward sections of combinatorial logic.

As simple as these devices are, they amply demonstrate the significant advantages that programmable logic provides. The increased reliability, decreased cost, and improved design time resulting from these devices did not go unnoticed. It was only natural to look for ways of expanding the role of programmable logic in the overall system design.

In this chapter we will look at some of the more sophisticated solutions designers have come up with for expanding the capability of programmable logic. Our discussion is not intended to be exhaustive. New and sophisticated devices are literally being announced weekly. Sometimes the announcements come daily. What we have done is to select key examples of the basic architectures of the advanced devices available. The devices discussed will provide a good understanding of each of the basic architectures. With this understanding, it is an easy task to look at a new device and weigh its relative merits. For all the ingenuity being shown in the development of PLDs, most new devices make use of the basic architectures discussed in this text.

In Chapter 4, we looked at how a PROM and a latch could be used to implement our baseline four bit counter. In Chapter 5 we saw how a registered PAL, the PAL16R4, could be used to realize the same function. Both implementations are reasonably efficient for our application. It is easy to see how these parts may not fit in as smoothly in other situations.

The state machine example is inherently inefficient since we did not need all of the product terms available in the used address space of the PROM (not to mention the fact that we only used a small fraction of the device, anyway). The pin allocation was not optimal either. As noted in the state machine example, if an additional seven output pins were available, the display driver could have been eliminated. We had plenty of extra input pins but they were of no use to us.

The situation was a little better with the PAL16R4 since we were able to use some of its combinatorial pins efficiently. And since we were trying to implement a four bit counter, its four D-registers worked out well. If we had needed a three bit counter, however, we would have been at a loss to make use of the remaining D-register. Since all of the registers are clocked in common, it is unlikely that we could have found a constructive way to employ a lone D-register. The register and all of its associated logic would simply have been wasted.

6.1 The PAL22V10

In an attempt to resolve the types of problems discussed above, AMD introduced the PAL22V10. The PAL22V10 was the first of the second generation PLD designed to overcome many limitations of the simpler first generation devices. The overall layout for the PAL22V10 is shown in Figure 6-1.

In Figure 6-1, we have introduced another PLD drawing convention, the *macrocell*. Macrocells are simply a "block diagram" approach to showing the overall architecture of a device. As the PLDs we discuss become more complicated, it is difficult and confusing to show all the details of the device. Therefore, the more complex devices are shown hierarchically, with the blocks detailed in a separate drawing.

The detailed view of the PAL22V10's logic macrocell is shown in Figure 6-2. Notice that the programmable AND array/fixed OR array follows the same basic pattern of the PALs that we have seen before. A significant point of interest is that the number of product terms in the macrocells varies. Like many PAL architecture devices, this *variable distribution* of product terms can affect implementation of a particular design. A design that will compile with one selection of pins may not compile with a different selection of pins.

The innovative part of the PAL22V10 is the output logic macrocell. The internal design of the PAL22V10's output macrocell is shown in Figure 6-2. The D-register in Figure 6-2 is similar to what we have already seen in the PAL16R4. Two interesting features have been added, however. First is an asynchronous reset on all registers. Setting the asyn-

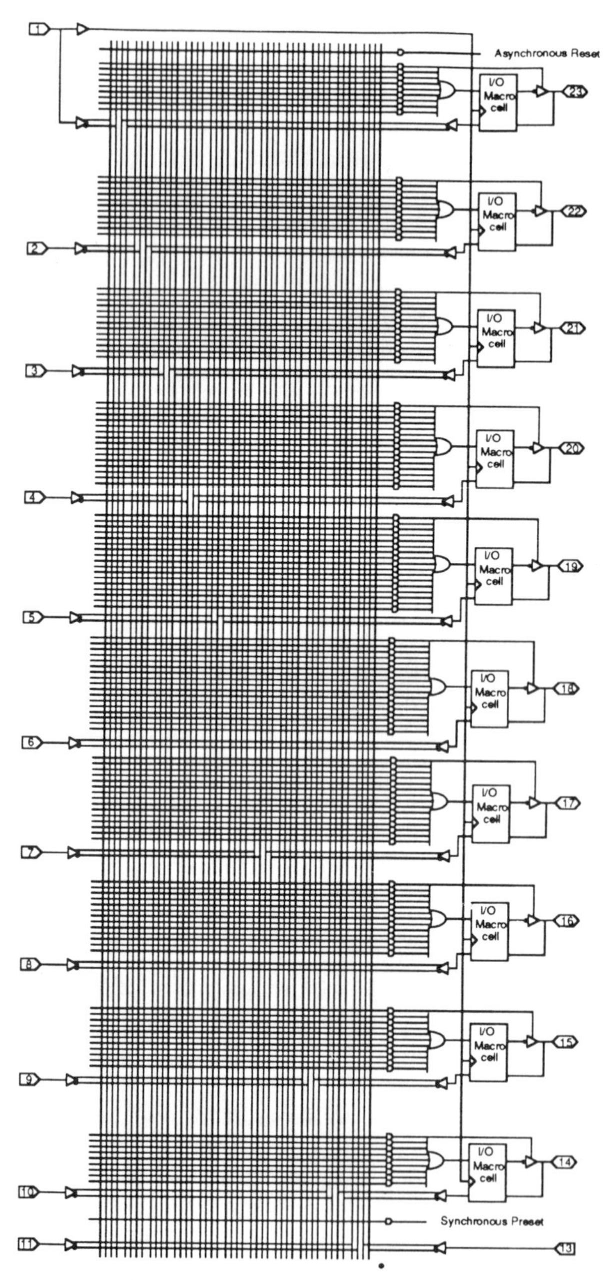

Figure 6-1
PAL22V10 Layout.

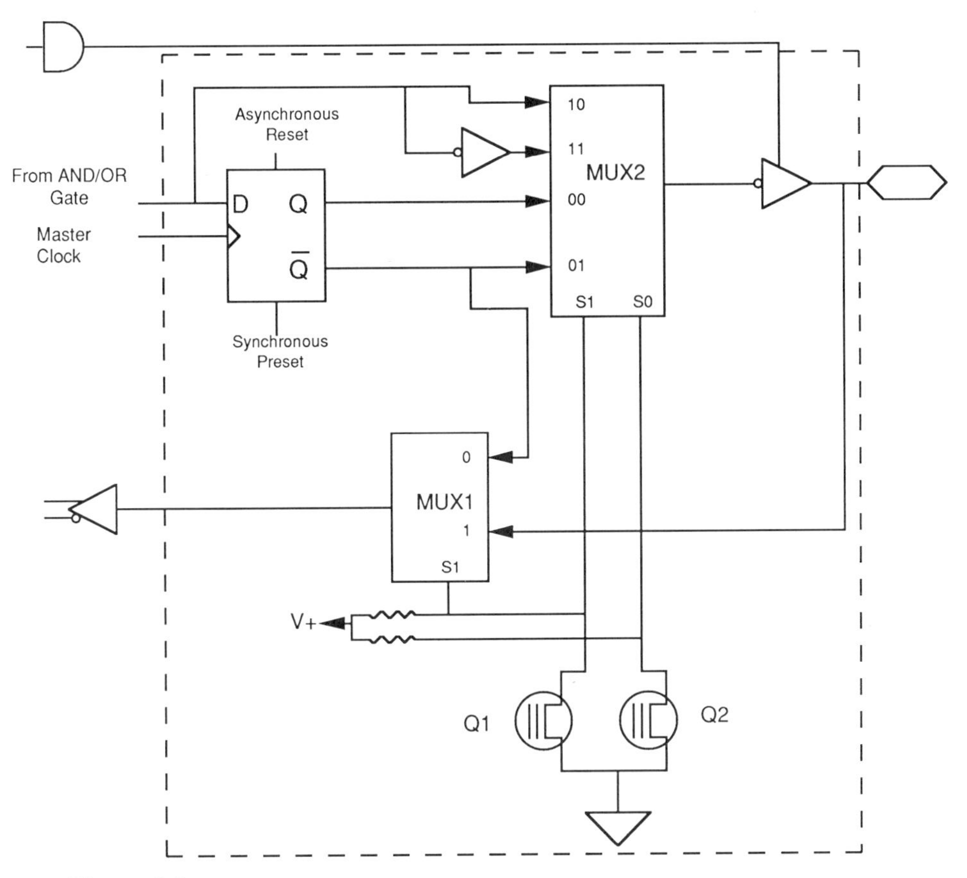

Figure 6-2
PAL22V10 Macrocell.

chronous reset will cause all of the D-registers to have a 0 on the Q output regardless of anything else that is happening. Setting the synchronous preset will cause all of the registers to have a 1 on the Q output *on the next clock transition*.

In addition to the register, the macrocell also contains two multiplexers: MUX1 and MUX2. These multiplexer blocks allow for a variety of configurations for the I/O pin.

Since there are two programming cells in the macrocell, Q1 and Q2, there are four basic configurations available:

Q1 ON, Q2 ON In this configuration, the output is a registered active low output. MUX1 routes the $/Q$ signal of the D-latch back to the

AND array. The output pin is driven via the inverting input to MUX2 (see Figure 6-3a).

Q1 ON, Q2 OFF This is similar to the previous case. The output pin is driven by the noninverting registered input to MUX2, however. This results in registered active high output (see Figure 6-3b).

Q1 OFF, Q2 ON In this case, MUX2 selects its input signal as the inverted input to the D-latch. This effectively bypasses the D-latch. At the same time, MUX1 selects its input from the I/O pin. The result is a combinatorial active low I/O pin (see Figure 6-3c).

Q1 OFF, Q2 OFF This is similar to the previous case. MUX2, however, selects its input from the noninverted input of the D-latch. Figure 6-3d shows the resulting combinatorial active high signal.

The I/O macrocell gives the PAL22V10 considerable flexibility. It allows the designer to select the number or registers needed for the particular application. Any unused output pins can be used for combinatorial functions. Overall, the designer can select from as many as 21 inputs and 1 output, to a more balanced 12 inputs and 10 outputs.

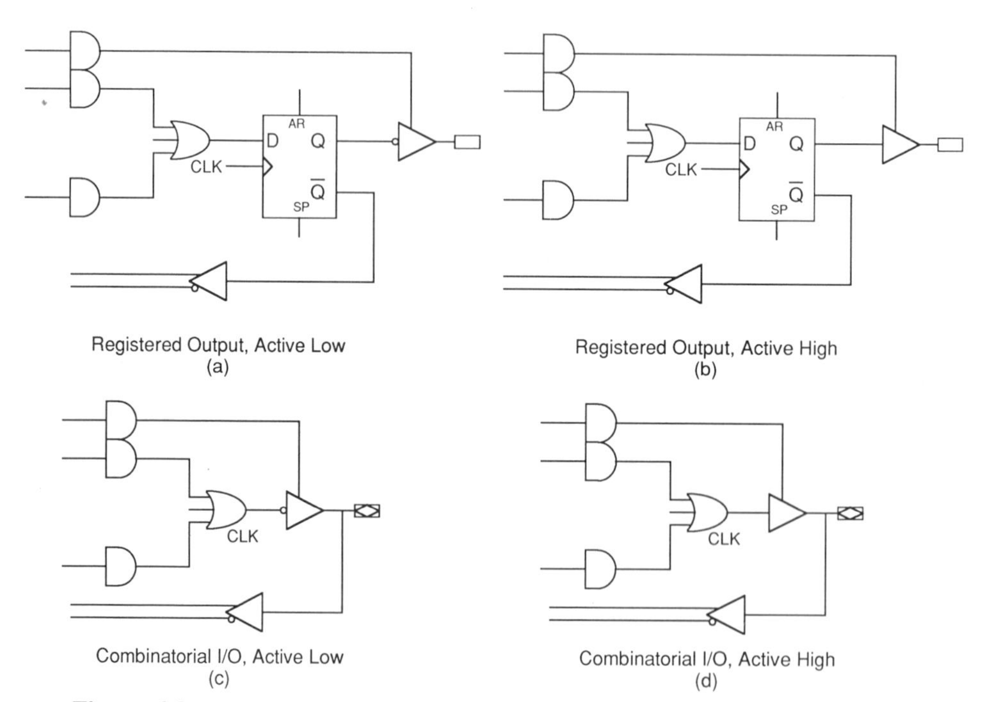

Figure 6-3
Possible I/O Combinations for the PAL22V10.

Another interesting feature of the PAL22V10 is that it is a PLD's PLD. It can be used to (functionally) replace other PLD devices. It is possible to use a PAL22V10 in place of such parts as the PAL16L8, the PAL16R4, and various other registered and combinatorial devices. The devices may not be pin-for-pin compatible, but this flexibility and the powerful functionality of the PAL22V10 makes it one of the most popular PAL devices available.

The ability of the PAL22V10 to replace other PALs can be seen by using it to implement our PAL-based four bit counter from Chapter 5. The PAL22V10 defaults to a registered active low output, so there are only two changes required to our source file.

First, naturally, we must change the include line from

```
include P16R4;
```

to

```
include P22V10;
```

The other change is to accommodate the programmable three-state drivers on the PAL22V10. The PAL16R4's three-state drives are hard-wired, so we did not need to worry about them in the programming. On the other hand, each of the PAL22V10's output drives is individually controlled by its own product term. We can add this control by simply including the following statements:

```
D3.oe = !OE;
D2.oe = !OE;
D1.oe = !OE;
D0.oe = !OE;
```

With these changes, the PAL22V10 would be functionally identical with our PAL16R4 implementation. Notice that it would not be pin-for-pin compatible, however, since the PAL16R4 is a 20 pin device and the PAL22V10 is a 24 pin device.

Since these changes are trivial to the overall theory of operation of the circuit, we have not duplicated the source listing with these minor changes.

Of more interest than simply mimicking the functionality of the PAL16R4 is the upgrading of the design. One of the disadvantages of this PAL16R4 design was that the reset was synchronous. In some applications this is not a problem, but for a push button counter it is somewhat awkward. In the next example, the asynchronous reset capability of the PAL22V10 will be used to include a more practical reset.

While we are at it, we will cover how to define which form of output each of the PAL22V10's I/O macrocells may take. The PAL22V10 can make use of either registered active low or registered active high outputs.

As the PAL16R4 example demonstrated, there is no theoretical difficulty in making use of a registered, active low output. However, for the PAL22V10 example a registered active high output is used. There are two reasons for this: First, our asynchronous reset sets the Q output of the D-latch to a zero. For the push button reset to clear the counter to zero, the output must be configured as an active high signal. Second, it is easier to understand the design and documentation if the positive form of logic is shown.

As noted earlier, the output macrocells default to registered, active low outputs. This means that if we do not tell the compiler (in some way) explicitly that another type of output is needed, then a registered active low device will result. The compiler is quite flexible, however, and will generally create the right output from the implicit information available.

For example, in forming the circuit for the RS debouncing latch the expression

```
!QOUT = SET & QBAR;
```

is used to form one of the NAND gates. Since this expression does not make use of the macrocells D-latch, the compiler will correctly configure the output macrocells as a combinatorial, active low signal.

The situation is different for the registered, active high output configuration. Notice that in the state table definition we have removed the "!" operator from the state variable declarations. This is not enough information, however, for the compiler to know that we want to use the registered, active high macrocell configuration. To give the compiler this information we must explicitly state:

```
D3 = q;
D2 = q;
D1 = q;
D0 = q;
```

These and a few other changes are shown in the listing presented in Figure 6-4.

Since the PAL22V10-based design makes use of the asynchronous reset, we remove the RESET declaration from the state definition. Notice that we have used two symbolic names: "RESET" and "reset." The reason for this is that proLogic recognizes "reset" as a predefined signal name. "RESET," on the other hand, is treated as a simple user variable

```
/****************************************************************************/
/*                       4 Bit Counter, 22V10 example.                      */
/* Function: 4 bit counter with asynchronous clear and three-state          */
/*           outputs.                                                       */
/* Device: 22V10.                                                           */
/*                                                                          */
/* Engineer: Jim Broesch                                                    */
/* Date: July 10, 1990                                                      */
/* This files is coded for compilation with proLogic.                       */
/****************************************************************************/
include p22v10;

/* Input pin definitions. */
define CLOCK = pin1;
define RESET = pin2;
define UP    = pin3;
define OE    = pin11;

/* Output pin definitions. */

define D3 = pin17;
define D2 = pin16;
define D1 = pin15;
define D0 = pin14;

/* Reset Latch  definitions. */

define SET   = pin4;
define CLEAR = pin5;

define QOUT = pin19;
define QBAR = pin18;

/* Configure I/O Macrocell as a registered, active high output. */
D3 = q;
D2 = q;
D1 = q;
D0 = q;

/* Put three-state drivers under control of OE (pin 11). */
D3.oe = !OE;
D2.oe = !OE;
D1.oe = !OE;
D0.oe = !OE;

reset = !RESET; /* Connect all flip-flop resets to RESET. */
!QOUT = SET & QBAR;    /* Build the RS     */
!QBAR = CLEAR & QOUT;  /*   Latch circuit. */

state_diagram D3,D2,D1,D0 {
     state s0 = 0000 {if (UP) s1; else sF;}
     state s1 = 0001 {if (UP) s2; else s0;}
     state s2 = 0010 {if (UP) s3; else s1;}
     state s3 = 0011 {if (UP) s4; else s2;}
     state s4 = 0100 {if (UP) s5; else s3;}
     state s5 = 0101 {if (UP) s6; else s4;}
     state s6 = 0110 {if (UP) s7; else s5;}
     state s7 = 0111 {if (UP) s8; else s6;}
     state s8 = 1000 {if (UP) s9; else s7;}
     state s9 = 1001 {if (UP) sA; else s8;}
     state sA = 1010 {if (UP) sB; else s9;}
     state sB = 1011 {if (UP) sC; else sA;}
     state sC = 1100 {if (UP) sD; else sB;}
     state sD = 1101 {if (UP) sE; else sC;}
     state sE = 1110 {if (UP) sF; else sD;}
     state sF = 1111 {if (UP) s0; else sE;}
                          }
```

Figure 6-4
PAL22V10 Four Bit Counter. (*Figure continues.*)

```
/* The following vectors perform an abbreviated functional test on the */
/*    counter.                                                          */

test_vectors {  /* Test the up/down counter logic. */
  CLOCK RESET UP OE  D3  D2  D1  D0   ;
    0     1    0  1   Z   Z   Z   Z   ;
    0     0    1  0   L   L   L   L   ;
    c     1    0  0   H   H   H   H   ;
    0     0    0  0   L   L   L   L   ;
    c     1    1  0   L   L   L   H   ;
    c     1    1  0   L   L   H   L   ;
    c     1    1  0   L   L   H   H   ;
    c     1    1  0   L   H   L   L   ;
    c     1    1  0   L   H   L   H   ;
    c     1    1  0   L   H   H   L   ;
    c     1    1  0   L   H   H   H   ;
    c     1    1  0   H   L   L   L   ;
    c     1    1  0   H   L   L   H   ;
    c     1    1  0   H   L   H   L   ;
    c     1    1  0   H   L   H   H   ;
    c     1    1  0   H   H   L   L   ;
    c     1    1  0   H   H   L   H   ;
    c     1    1  0   H   H   H   L   ;
    c     1    1  0   H   H   H   H   ;
    c     1    1  0   L   L   L   L   ;
    c     1    0  0   H   H   H   H   ;
    c     1    0  0   H   H   H   L   ;
    c     1    0  0   H   H   L   H   ;
    c     1    0  0   H   H   L   L   ;
    c     1    0  0   H   L   H   H   ;
    c     1    0  0   H   L   H   L   ;
    c     1    0  0   H   L   L   H   ;
    c     1    0  0   H   L   L   L   ;
    c     1    0  0   L   H   H   H   ;
    c     1    0  0   L   H   H   L   ;
    c     1    0  0   L   H   L   H   ;
    c     1    0  0   L   H   L   L   ;
    c     1    0  0   L   L   H   H   ;
    c     1    0  0   L   L   H   L   ;
    c     1    0  0   L   L   L   H   ;
    c     1    0  0   L   L   L   L   ;
               }

/* Test vectors for the RS Latch. */

test_vectors{  /* Test the RS switch debouncing latch. */
     CLEAR SET QOUT QBAR ;
       0    1   L    H   ;
       1    1   L    H   ;
       1    0   H    L   ;
       1    1   H    L   ;
           }
```

Figure 6-4
(*Continued*)

and would not directly connect to the reset input of the D-registers. A schematic would be similar to the one shown in Figure 5-7, so we will not replicate the schematic again here.

While the specific coding examples shown here apply to proLogic, the same basic considerations must be made for whatever package is used for development.

Finally, the importance of the flexibility provided by the PAL22V10 should be emphasized. While earlier parts had made efforts to be as flexible as possible, they did not attain the degree of acceptance that the

PAL22V10 achieved. This set the stage for the more sophisticated parts available today. These include the GAL series of devices and the popular EPM-310 from Altera. These devices offer greater flexibility and even more sophistication than the PAL22V10. The PAL22V10 has not lost its following, however, and is one of the most popular PLDs available.

The PAL22V10 is second-sourced by a large number of manufacturers and can be found in both bipolar and CMOS technologies. In general, the CMOS versions are based on an EPROM technology and can be erased by exposure to ultraviolet light.

6.2 PSG-506/507

Texas Instruments PSG-506 and PSG-507 sequencers are particularly interesting parts. This style of device was originally introduced by Signetics in its popular PLHS105. The PLHS105 is still around and commonly used. It may be an attractive option when a simpler state machine is required. For this discussion, however, we will look at the more flexible and powerful PSG-506.

The PSG-506 combines a large PLA architecture with a set of "buried" D-flip-flops forming the pipeline register. The term "buried" means that there is no direct connection from the pins of the IC to the flip-flops. A simplified block diagram of the PSG-506 is shown in Figure 6-5.

Architecturally the PSG-506 and PSG-507 are PLA-based circuits. The output of the AND array is divided into two parts. The first part is feedback to the AND array. Most of the feedback is done via RS registers, however, and two complemented product terms are fed back directly to the array. The feedback registers form a pipeline register that implements a classic state machine architecture. Either a Mealy or Moore state machine can be implanted. The second part of the AND array is fed, via the I/O control blocks, to the output pins.

The PSG-507 is similar to the PSG-506 but has an internal six bit counter as part of the pipeline register. Since many timing and control functions make use of sequential states, the use of a counter as part of the pipeline is more efficient in these applications. The counter allows product terms in the array to be freed for use in complex state transitions. Otherwise, a large number of product terms would be required to perform the sequential counting task. In both devices the use of buried registers is attractive since it frees pins for other functions.

The I/O control blocks can be configured as either a combinatorial or registered output. Unlike the PAL22V10, the outputs are positive only and cannot be configured as inverting outputs. The clock is sent to all of the registers (including the output registers) via a programmable inverter.

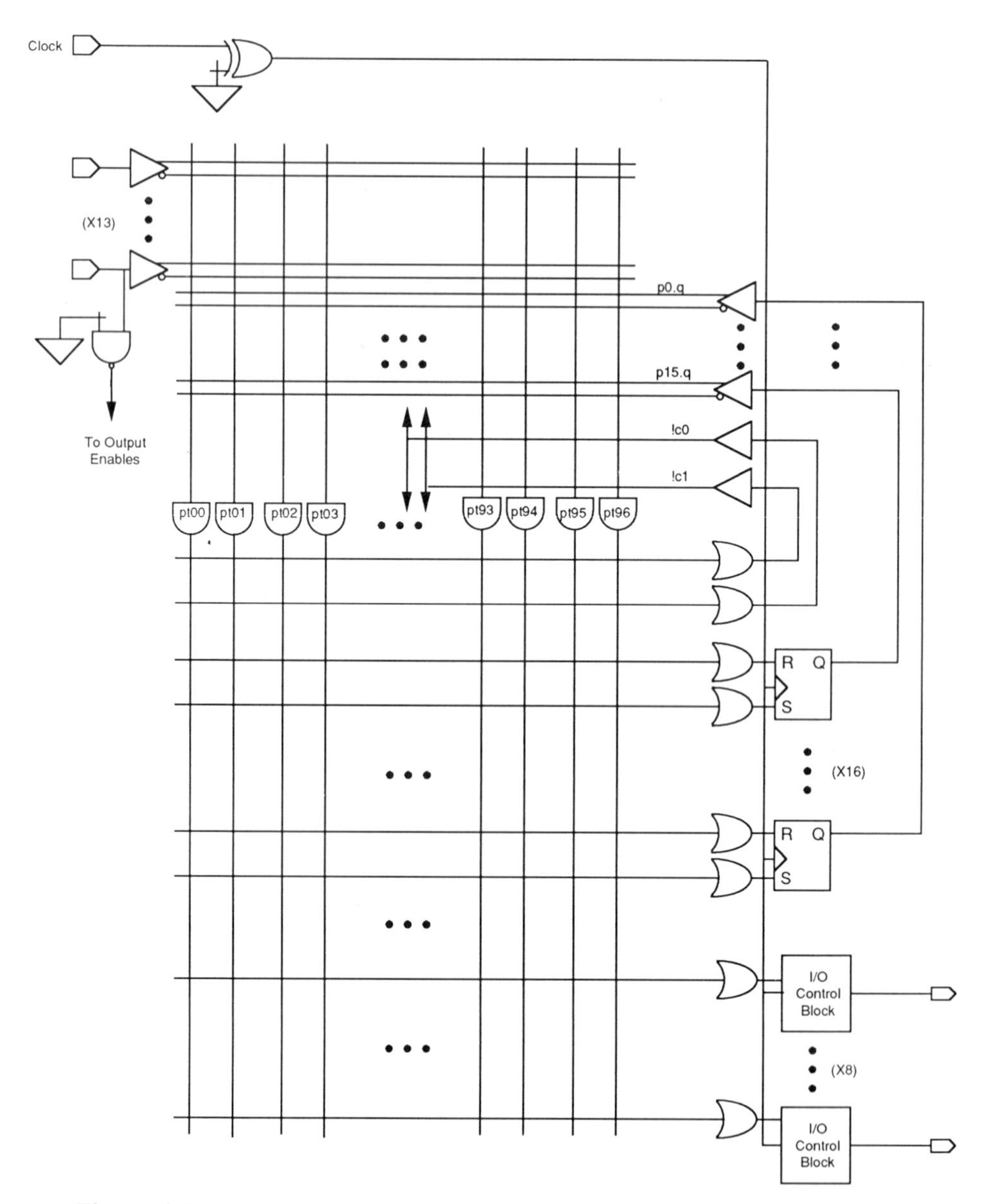

Figure 6-5
Simplified Diagram of PSG-506.

This allows the PSG-506 to be clocked on either edge of the clock. The outputs can be configured as permanently enabled or the three-state drivers can be controlled from the pin 17 input. The architecture of the PSG-506 output control cell is shown in Figure 6-6.

As the name implies, the PSG-506/507 is designed primarily to generate timing sequences. Typical applications include dynamic memory controllers and bus arbiters. This device is also useful in syndrome generators, error detection and correction circuitry, and other such applications.

We will demonstrate the PSG-506's flexibility and show how a compiler can be used to integrate diverse functions by using a PSG-506 to realize our programmable counter in a single device. The debouncing RS latch, the counter, and the display decoder can all be accommodated by the PSG-506 architecture.

In the counter examples so far, we have used the current state as the output. We have then decoded the state with an external display decoder. This partitioning is a natural one and is often seen in designs making use of PLDs. On the other hand, it is always desirable to limit the number of components required to implement a function. TI's PSG-506/507, with its buried registers and decoded I/Os, allows us to do exactly this.

The schematic for the circuit is shown in Figure 6-7. Like many other PLD schematics, this one is not particularly explanatory. The real understanding of the circuit is gained from studying the source file shown in Figure 6-8. The definition section of the source file shown in Figure 6-8 is similar to the definition sections we have seen previously and requires no further discussion.

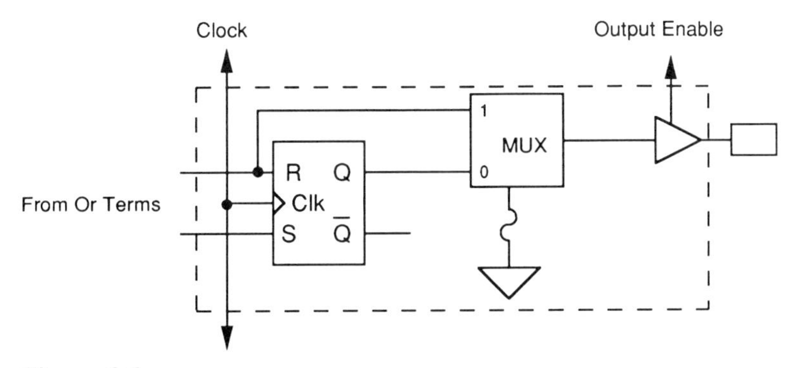

Figure 6-6
PSG-506 Output Control Cell. (Reprinted with permission of Texas Instruments, ©1990.)

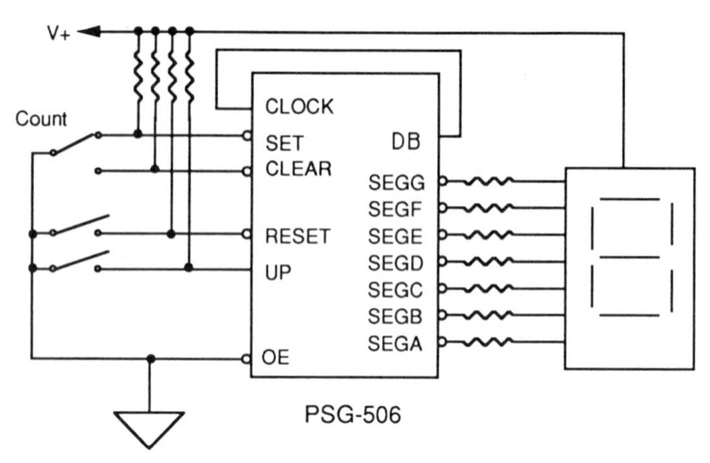

Figure 6-7
PSG-506 Version of the Four Bit Counter.

The first equation, oe = !pin17, simply puts the three-state buffers of the output cells under control of pin 17. This is not particularly necessary for our design since the outputs are always enabled. We do not need the input pin, however, so we might as well put it to use. Being able to put the outputs in a high impedance state can be useful in production testing or sometimes even in device testing. This situation often arises when using PLDs; excess capabilities exist so they may as well be exploited. Even if these capabilities are not needed in later parts of the product life cycle, there is little extra cost in making use of the excess capability. We will look at this more closely in the next chapter when designing for testability is discussed.

The RS latch equations are more interesting, particularly because we make use of the internal asynchronous feedbacks of the PSG-506 to implement our latch. In previous examples we were able to directly implement this circuit. In this case, however, we are more limited. This comes about for two reasons: First, since we are using seven of the eight output pins for drive signals, we have only one output left for the latch. Second, the lack of inverting outputs limits the configuration options. To make the presentation of the equations clearer, we have shown an equivalent circuit in Figure 6-9.

First we connect the feedback terms, c0 and c1, to product terms pt0 and pt1:

```
c0 = pt1;
c1 = pt0;
```

```
/*************************************************************/
/*                  PSG Based 4 Bit counter.                 */
/* Device: TI PSG-506.                                       */
/* Function: Implements a 4 bit counter with                 */
/*           asynchronous preclear and built in               */
/*           seven segment LED decoder.                      */
/*                                                           */
/* Author: Jim Broesch                                       */
/* Date: August 2, 1990                                      */
/*************************************************************/
include A506;

/* Input pin definitions. */

define CLOCK = pin1;
define RESET = pin2;
define UP    = pin3;
define SET   = pin4;
define CLEAR = pin5;
define OE    = pin17;

/* State variable name assignments. */

define D3 =  p3.q;
define D2 =  p2.q;
define D1 =  p1.q;
define D0 =  p0.q;

/* Internal node names. */

pt03 = D3;
pt02 = D2;
pt01 = D1;
pt00 = D0;

/* Map internal node names to output pins. */

define OE = pin17;

define G = pin15;
define F = pin14;
define E = pin13;
define D = pin11;
define C = pin10;
define B = pin9;
define A = pin8;

/* Next comes the logic equations. */

oe = !OE; /* Connect the output enables. */

pt04 = SET & !c1;    /* RS Latch    */
pt05 = CLEAR & !c0;  /*   circuit.  */

c0 = pt04; /* Feed back the cross    */
c1 = pt05; /*  coupled terms.        *

pin16 = pt05; /* Connect one side of the latch to */
              /*    the outside world.            */

state_diagram D3,D2,D1,D0 { /* Define the states. */
     if (pin2) s0;
     state s0 = 0000 {if (UP) s1; else sF;}
     state s1 = 0001 {if (UP) s2; else s0;}
     state s2 = 0010 {if (UP) s3; else s1;}
     state s3 = 0011 {if (UP) s4; else s2;}
     state s4 = 0100 {if (UP) s5; else s3;}
     state s5 = 0101 {if (UP) s6; else s4;}
     state s6 = 0110 {if (UP) s7; else s5;}
     state s7 = 0111 {if (UP) s8; else s6;}
     state s8 = 1000 {if (UP) s9; else s7;}
     state s9 = 1001 {if (UP) sA; else s8;}
     state sA = 1010 {if (UP) sB; else s9;}
     state sB = 1011 {if (UP) sC; else sA;}
     state sC = 1100 {if (UP) sD; else sB;}
     state sD = 1101 {if (UP) sE; else sC;}
     state sE = 1110 {if (UP) sF; else sD;}
     state sF = 1111 {if (UP) s0; else sE;}
                     }
```

Figure 6-8

PSG-506-Based Counter. (*Figure continues.*)

```
/* Now comes the new stuff. Notice that the internally named */
/*  states (D0,D1,D2, and D3) are input to the truth table.  */
/* The output of the truth table is the output pins.         */

truth_table{
     D3 D2 D1 D0  :   G  F  E  D  C  B  A  ;
     0  0  0  0   :   0  0  0  0  0  0  0  ;   /* 0 */
     0  0  0  1   :   1  1  1  0  0  1  1  ;   /* 1 */
     0  0  1  0   :   0  0  0  1  0  0  1  ;   /* 2 */
     0  0  1  1   :   0  1  0  0  0  0  1  ;   /* 3 */
     0  1  0  0   :   0  1  1  0  0  1  0  ;   /* 4 */
     0  1  0  1   :   0  1  0  0  1  0  0  ;   /* 5 */
     0  1  1  0   :   0  0  0  0  1  0  0  ;   /* 6 */
     0  1  1  1   :   1  1  1  0  0  0  1  ;   /* 7 */
     1  0  0  0   :   0  0  0  0  0  0  0  ;   /* 8 */
     1  0  0  1   :   1  1  0  0  0  0  0  ;   /* 9 */
     1  0  1  0   :   0  1  0  0  0  0  0  ;   /* A */
     1  0  1  1   :   0  0  0  0  1  1  0  ;   /* B */
     1  1  0  0   :   1  0  0  1  1  0  0  ;   /* C */
     1  1  0  1   :   0  0  0  0  0  1  1  ;   /* D */
     1  1  1  0   :   0  0  0  1  1  0  0  ;   /* E */
     1  1  1  1   :   0  0  1  1  1  0  0  ;   /* F */
           }
```

Figure 6-8
(*Continued*)

Next, we cross-couple the outputs and link in the external inputs:

```
pt01 = SET & !c1;
pt00 = CLEAR & !c0;
```

Finally, we need to connect the output pin 16 (which we named "DB") back to the latch. This requires a bit of navigating to show the compiler what we want:

```
pt2 = !c0;
```

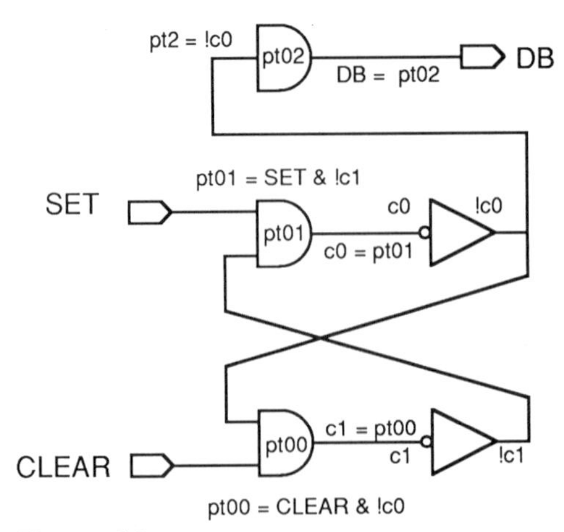

Figure 6-9
RS Latch Using the PSG-506 Architecture.

This creates a *parallel* path from the !c0 signal through the product terms and into the OR array. We could not simply tap the same product term used in the NAND gate; the location of the INVERTER in the feedback terms does not allow it.

Next, we need to get the output of the latch, now buffered by pt02, to pin 16. We do this with:

```
DB = pt02;
```

This sequence is best understood by comparing the architectural view of the PSG-506, the listing in Figure 6-7, and the schematic in Figure 6-8. The reason that it is worth the trouble to understand this process is that it is a good illustration of how one must navigate around a complex PLD. A good compiler like proLogic will relieve us of much of the burden of programming a PLD. Ultimately, however, the designer is responsible for understanding the target device well enough to provide the layout information. A lack of appreciation for this fact is one of the major roadblocks that face designers who are new to using PLDs. This is particularly true of designers who have a strong hardware background but little exposure to software development.

The rest of the software example is more straightforward. The state diagram is virtually identical to those we have seen before. This is also true of the truth table. The interesting thing is the way that proLogic and the PSG-506's architecture allow us to combine the two distinct forms in single file.

We could have incorporated the decoding directly into the state diagram. In many cases this is exactly what we would like to do. In this case, however, the decoding table is sufficiently complex that including the decoding in the state diagram would have been quite cumbersome.

In keeping the state diagram and the truth table separate, we make it very easy to see exactly what each is doing. This aids greatly both in the documentation of the design and in debugging the code while we are developing the basic design.

Finally, we will conclude with some general observations and comments on the PSG-506 and PSG-507. These devices are quite powerful and relatively flexible, but the fact that they are designed for a specific purpose should not be overlooked. Their major purpose is to provide timing waveforms. The actual waveforms may vary from direct digital synthesis to dynamic memory control, but the basic architecture is optimized for this type of circuit. We emphasize this point to put the upcoming observations in perspective: they are not criticisms, rather they simply reflect the different focus of the device.

The 506 architecture suffers from many of the same limitations as found in the PROM architecture. The number of inputs are fixed, as are the number of outputs. No mixing and matching is possible, as with devices such as the PAL16L8 or the PAL22V10. Furthermore, while the 506 can be used for purely combinatorial functions, it does not handle this role well. Finally, all of the registers in the device are clocked by a common clock. As with the 16R series and the PAL22V10, this limits the use of these devices to synchronous designs.

6.3 ASICs and Third Generation PLDs

In order to understand the architecture of the next two devices and almost as importantly their supporting software, it is necessary to first understand some of the philosophy that led to their introduction. In the early 1980s, PLDs were used primarily to replace awkward sections of combinatorial logic. Principally, this meant things such as address decoders and parity generators. Simultaneously, the idea of the application specific integrated circuit (ASIC) was first being introduced. The acronym "ASIC," like many of the acronyms used in electronics, has come to mean many different things to many different people.

In this book we use the term ASIC with its original meaning: an IC that was designed to be used in a specific application for a customer. As such, the term implies that the IC was developed for, or by, the customer as either a custom device or a gate array. We do not consider microprocessors, RAMS, standard peripherals, or other such commodity devices to be ASICs.

Custom ICs come in essentially two flavors: full-custom or semicustom. As the name implies, full-custom ICs are essentially designed from scratch. Semicustom ICs are also designed from scratch but they have predefined structures such as gates and flip-flops already in the layout library.

Unlike custom ICs, gate arrays are composed of a large *array of gates*. The gates are literally wired together by metallization layers of the IC.

While it is somewhat of an oversimplification, the basic trade-offs can be stated as follows:

- Custom ICs cost the most to design and get working. But once they are working, they are the cheapest of the options to produce since they fit the most circuitry on the least silicon area. For the same reasons, custom ICs will usually have the highest performance.
- Semicustom ICs are the second most expensive to get working but are nearly as economical to produce as full-custom ICs. They also provide high-performance.

- Gate arrays are the cheapest to design with, since the basic building blocks are already in place. Since gate arrays can go through the expensive diffusion processes prior to metallization, they can be produced in bulk. When a user needs them, the metallization can be added to realize the customer's needs. On the other hand, gate arrays often cannot make full use of the gates on a device since these gates are laid out prior to the design of the ICs. The metallization layer allows only limited connections.

The above discussion assumes an "everything else being equal" scenario. In practice, the actual cost and performance benefits require careful evaluation. A poorly designed custom IC may well cost more per chip and have a lower performance than a well designed gate array.

The relevance of this discussion to PLDs is twofold. First, engineering decisions are often made on the basis of trade-offs between PLDs and ASICs. Second, PLD manufacturers are continuously trying to push the trade in favor of PLDs.

This brings us to the discussion of the following two architectures: Altera's MAX architecture and Xilinx's RAM based FPGA architecture. Both families of devices have the same basic objective: to raise the sophistication of the PLD to the point that they can effectively compete with gate arrays and possibly even custom ICs.

There are several significant advantages in using PLDs over ASICs. Even if the PLD's cost per device is not competitive with ASICs, other factors will generally make the overall cost of a PLD much lower. One of the biggest factors is time. Any PLD can be programmed in the system the same day the design is finished. A gate array may take weeks or months to get back, and a full-custom design may take a year or more. Obviously, this time to market versus per device cost must be weighed significantly.

This is particularly true when one considers that as many as half the custom IC designs do not work in the target system the first time they are plugged in. This failure is usually not caused by design of the IC but rather because of design problems at the system level. Still, to correct the problem means weeks or months added onto the design schedule.

It was with these factors in mind that PLD designers set out to produce the advanced PLDs we will be discussing next.

6.4 Altera

Altera decided to produce an advanced PLD basically by combining many PALs into one large package. They first did this with their EP-310 through EP-1800 series of devices. The EP-310 was basically a sophisticated PAL22V10. While the EP-310 has many significant features over the PAL22V10, its basic architecture is not sufficiently unique to warrant discussion.

The innovative aspect of the EP family is that the higher integration devices basically used multiple EP-310s connected together. This allowed one device to replace collections of PLDs. The EP family is quite popular, and Altera is currently licensing the design to other vendors.

Following the success of the original EP family, Altera introduced the multiple array matrix (MAX) family of devices. The MAX family is sufficiently unique to warrant closer attention.

Like the basic EPs, the MAX devices are modular and based on a PAL-like structure. Figure 6-10 shows the basic modular layout of the MAX family. We will begin our discussion of the MAX family by looking at the basic building block, the macrocell. The macrocell is shown in Figure 6-11. There are several key points to notice about the macrocell:

- The D-register is included as part of the macrocell.
- The clock for the D-register can come from either the chip-wide master clock or the logic array. This means that both synchronous and asynchronous state machine designs are realizable.
- There are only three product terms summed by the OR gate. More on this later.
- The output of the OR gate goes to a programmable inverter controlled from the logic array.

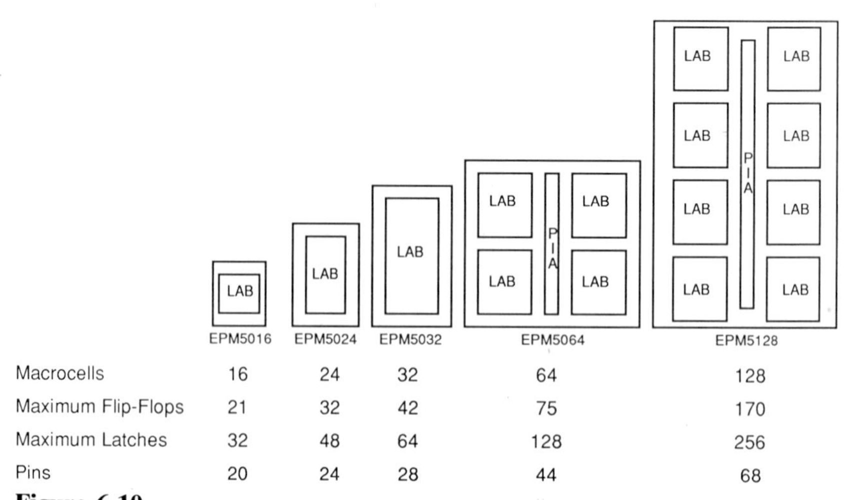

	EPM5016	EPM5024	EPM5032	EPM5064	EPM5128
Macrocells	16	24	32	64	128
Maximum Flip-Flops	21	32	42	75	170
Maximum Latches	32	48	64	128	256
Pins	20	24	28	44	68

Figure 6-10
MAX Family Architecture. (Reprinted courtesy of Altera Corporation, ©1990.)

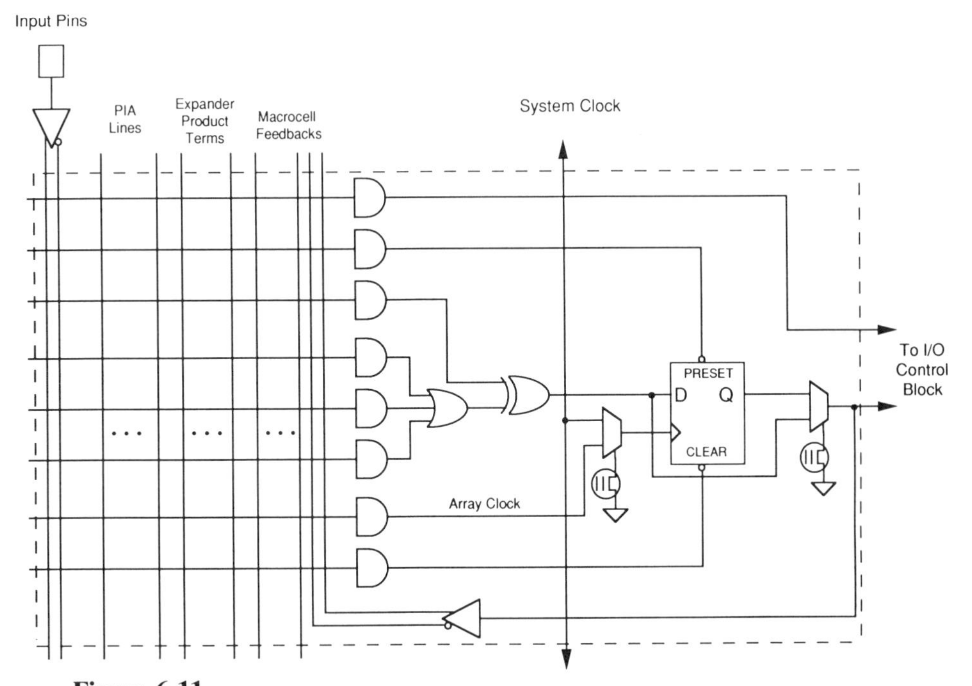

Figure 6-11
MAX Logic Macrocell. (Reprinted courtesy of Altera Corporation, ©1990.)

- Like the PAL22V10 architecture, the D-registers can be bypassed. Thus both registered and combinatorial designs can be realized.
- Looking at the logic array, notice that there are two additional groups of signals. Not only are the familiar input and feedback signals in the array, but there is also a grouping for expander product terms and one for the programmable interconnect matrix (PIA).
- The output of the D-register is fed back to the logic array. Notice that this occurs *before* the signal is routed to the I/O control block. The significance of this is that the register can be used as a buried logic element *independent* of the I/O pin.

One of the most interesting things about the MAX architecture is that it straddles the dividing line between a PAL and a PLA. As we noted earlier, 70% of PAL applications require three or fewer product terms. Most of the complexity of PALs such as the PAL22V10 is aimed at meeting the other 30% of the applications that do require more than three terms.

For a PLA, the 70% does not really mean much. The PLA architecture allows product terms to be distributed at will. The cost of this flexibility is, of course, the extra OR logic array.

The MAX architecture strives to achieve the best of both worlds. The simple three product term PAL architecture will meet the majority of programmable function needs. When more product terms are required, they can be added by using the "expander product terms."

Expander product terms are best understood by looking at the simplified drawing in Figure 6-12. The expander terms have both their inputs *and* outputs tied into the logic array. This arrangement allows the expander terms to be incorporated into very complex functions. Also notice that the product terms can be distributed among the macrocells or can be shared by several macrocells if the product term is common to each macrocell. This distribution of product terms is nearly as powerful as that

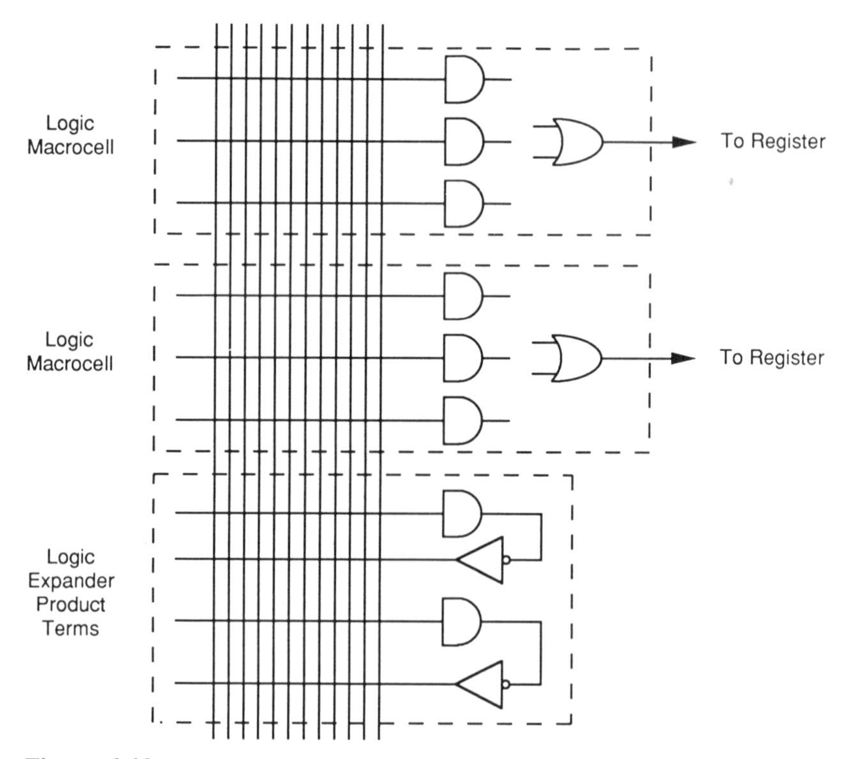

Figure 6-12
Simplified MAX Architecture Showing Expander Product Terms. (Reprinted courtesy of Altera Corporation, ©1990.)

found in a PLA. No extra overhead for an additional OR array is required, however.

Another interesting and useful feature of this architecture is that the expander product terms can be configured as extra latches or flip-flops. This allows wider counters, increased states, or more complex sequential functions to be realized than would be assumed by simply counting the registers in the architecture.

The next interesting feature of the MAX architecture is the I/O control block shown in Figure 6-13. This configuration is particularly interesting since it allows the output of the macrocell to be fed back into the logic array independently of the I/O pin. The significance of this is twofold: First, the registers can be *either* buried or visible; something none of the other devices we have looked at so far allow. Second, even if the I/O pin is used as an input, the logic macrocell can still be used as a buried register. This was one of the limitations we noted in the PAL22V10: if the I/O pin was used as an input, there was no practical way to make use of the associated macrocell.

The logic macrocell, the I/O control block, and the expander terms are combined to form what Altera calls the logic array block (LAB). The layout of the LAB is shown in Figure 6-14. LABs are connected via the programmable interconnect array. The overall architecture of the top of the line EPM5128 is shown in Figure 6-15.

The simpler members of the MAX family contain only a single LAB. These devices can be used in much the same way as the PAL22V10 or PSG-506. The high-end parts such as EPM5128 are quite powerful and are serious competition for even moderate-sized gate arrays. Since anything but a trivial practical example would be impossible to fit in the confines of these pages, we will defer from the usual practice of showing practical

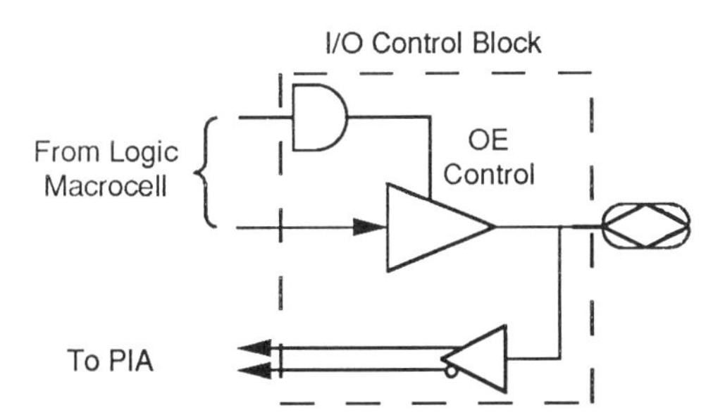

Figure 6-13
Detailed Drawing of the MAX I/O Control Block. (Reprinted courtesy of Altera Corporation, ©1990.)

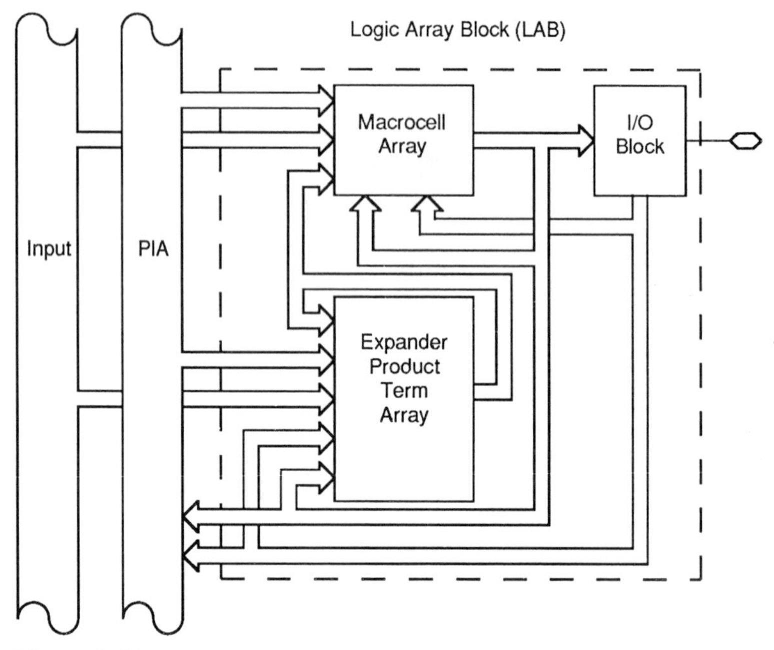

Figure 6-14
MAX Logic Array Block. (Reprinted courtesy of Altera Corporation, ©1990.)

circuits. Instead, some of the areas where these devices are used will be discussed along with practical considerations in employing them.

The first point of interest is that the MAX devices are based on EPROM technology. This means that these devices can be programmed, inserted into the circuit, and tested. If the design does not work, the part is simply erased and reprogrammed with the modified design. This flexibility and quick turnaround may reduce development costs by an order of magnitude over a gate array–based approach.

All of this power does not come without strings, however. One of these is the need to learn to work with a new software environment. The MAX family requires a special set of software to effectively support its unique features. Whether or not this is an issue depends on whether one is comparing the MAX devices to other PLDs or to gate arrays.

Gate arrays require their own special programming environments; so in comparison to gate arrays, the software is not an issue. In comparison to other PLDs, however, the unique MAX software locks one into the family. Unlike simpler devices such as the PAL22V10, there is not a wide range of PLD compilers available for use in the programming process. If

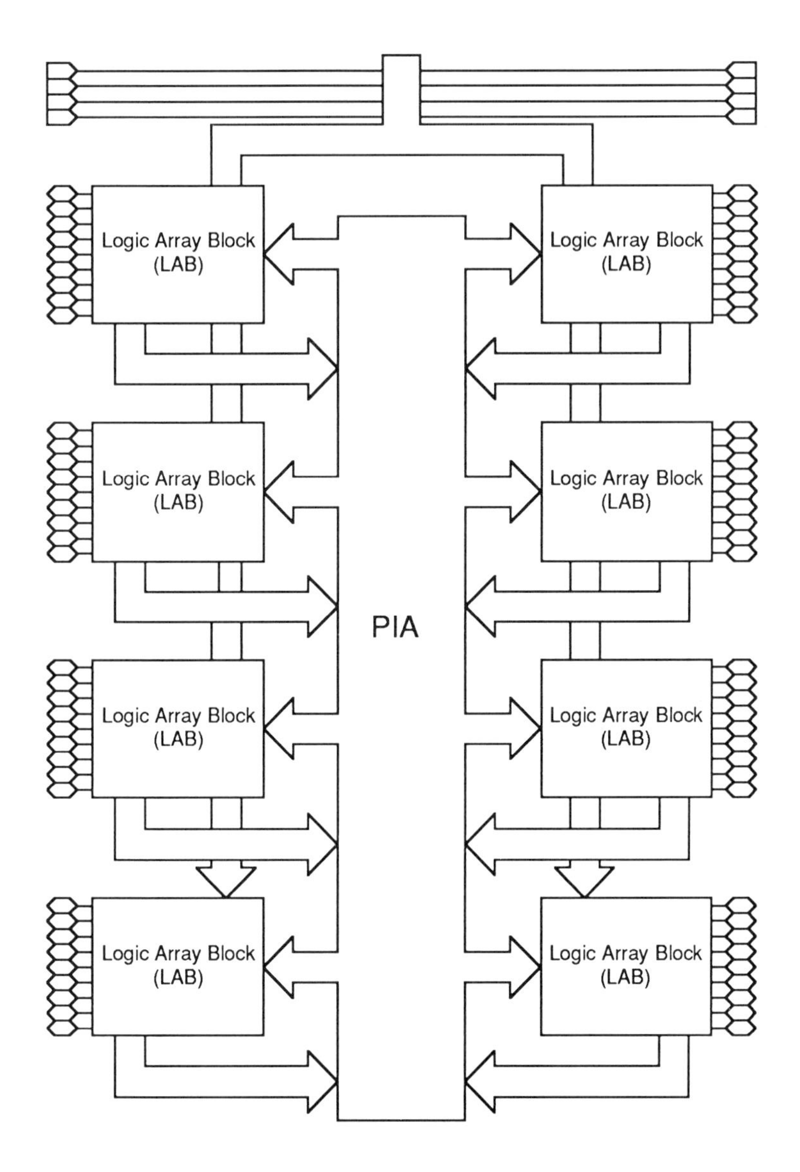

Note: 8 Input Pins, 52 Bidirectional Pins
16 Macrocells per LAB, 128 Macrocells Total
256 Total Expander Terms

Figure 6-15
EPM5128 Block Diagram. (Reprinted courtesy of Altera Corporation, ©1990.)

several other types of PLDs are also used in the design, this mix of programming vehicles complicates the design and documentation process.

The MAX family makes use of a very sophisticated development software called MAX+PLUS that employs both graphical and textual programming techniques. As noted in Chapter 5, a popular use of PLDs is to replace existing TTL-based logic. MAX+PLUS provides a library of equivalent 7400 series parts that can make the conversion from a TTL-based design to a MAX PLD almost painless.

The MAX device has gained wide acceptance in the industry. While it is unlikely to become *the* standard architecture, it is certainly going to be *one of the* industry standard architectures.

6.5 Xilinx

Xilinx products, known generally as field programmable gate arrays (FPGA), have taken a different and somewhat unique approach to programmable logic architectures. As devices previously discussed have shown, most PLDs take one of the three basic forms: PAL, PLA, or PROM. The FPGA architecture, on the other hand, is more of a true gate array.

The basic architecture of the FPGA is shown in Figure 6-16. There are two basic type of cells, called "blocks." The I/O block, as the name implies, is used to make the connections between the internal logic elements and the external world. The detailed diagram of the I/O block is shown in Figure 6-17. Notice that the I/O block has a D-register on the *input*. The D-register can be bypassed thus allowing for either a combinatorial or registered input. This registered input is particularly useful for synchronizing inputs for state machine designs or for interfacing the FPGA directly to conventional data buses.

The other major portion of the FPGA is the configurable logic block (CLB). The CLBs can be viewed as mini-registered PLDs. The basic layout of the CLB is shown in Figure 6-18. Between the various multiplexers, CLBs, and registers, a wide variety of functional configurations are available. These include:

- Any function of four variables.
- Two independent functions of three variables.
- Dynamic selection of two independent functions of three variables.
- Variations on the above themes, with the D-register being either an edge-sensitive or level-sensitive device.

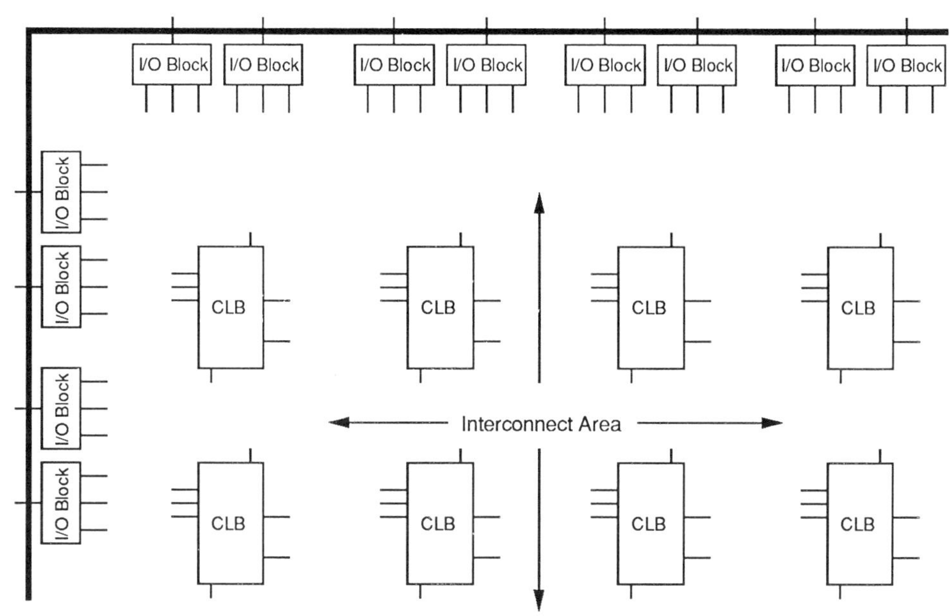

Figure 6-16
Field Programmable Gate Array Architecture (FPGA). (Reprinted courtesy of Xilinx, Inc.)

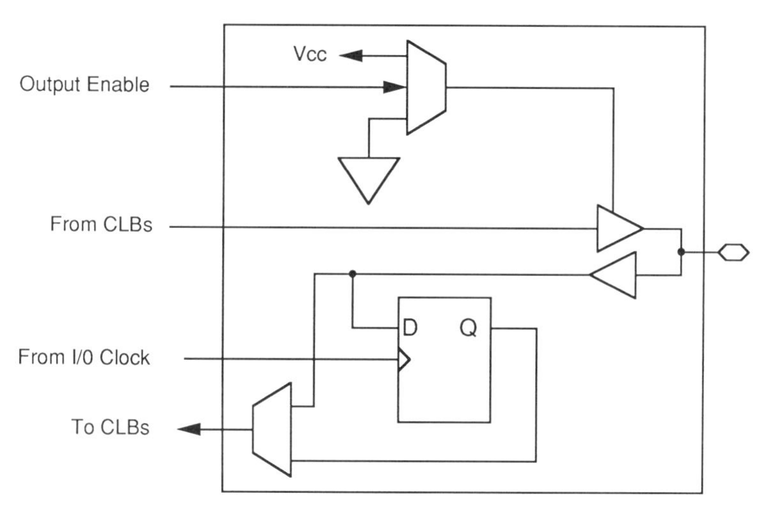

Figure 6-17
Detail of the I/O Block. (Reprinted courtesy of Xilinx, Inc.)

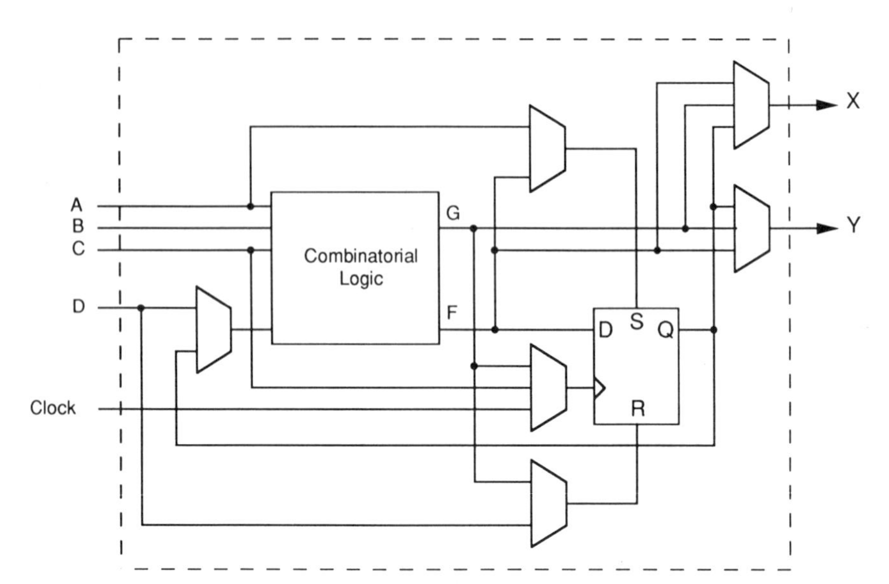

Figure 6-18
Configurable Logic Block (CLB). (Reprinted courtesy of Xilinx, Inc.)

The overall logic functionality of the device is defined by two factors: First, of course, is the selection of the logic function in the CLB; and second is the interconnection of the CLBs.

As is in a conventional gate array, the CLBs must be wired together. In a conventional gate array this is done by adding layers of metallization to the chip. In the FPGA, the connections are under the control of a memory cell similar to those found in conventional static RAMs. The actual connection paths are divided into three basic categories:

1. *Direct Connection.* These connections can be made between CLBs that are physically close to each other.
2. *Programmable Interconnects.* Programmable interconnects are routed via local switchboards called switching matrixes.
3. *Long Lines.* Essentially long lines are highways around the chip that allow for widely distributed CLBs to be connected together with minimal routing delays.

As with conventional gate arrays, it may not be possible to fully route a FPGA. The routing issue also extends to timing. The AC performance of the design will be a function of how the individual CLBs are connected together. The impact of this on the designer is that he or she must be concerned not only with the logical functionality of the design but also with how the design is "packaged."

As with the Altera devices, a practical example using a FPGA is beyond the scope of this book. Instead, some of the unique features of the FPGA are presented. In the next section we will make some comparisons between the FPGA and the MAX architecture.

One of the more interesting features of the FPGA is the fact that it is RAM-based. Most PLDs are based on either fuses, EPROM, or EEPROM technologies. These technologies are popular since the PLD, once programmed, retains its programming indefinitely. The RAM-based design of the FPGA is inherently volatile, and once the power is removed from the FPGA, all programming information is lost. This leads to some interesting options and considerations when using these devices.

Since the FPGA loses all of its programming when the power is removed, some means must be provided to reprogram the device at the start of each power-on cycle. This may be accomplished in several ways. One option is simply to provide a special external PROM that is specifically designed for loading FPGAs. These configuration PROMs have a serial architecture that minimizes board area and the number of interconnections necessary in an application. Or if the system has program memory, as in microprocessor based designs, the FPGA can be configured to get its information directly from system PROMs.

Perhaps the most direct solution is to simply not turn off the power to the FPGA. This can be accomplished with special battery backup circuits similar to those used in static memory backups.

Of course, all of these options involve some additional overhead in terms of the design time and utilization of available system resources. Whether or not this is really a factor in selecting a FPGA architecture depends on the particular application.

One of the themes of this book is to encourage the use of programmable logic to reduce parts count; this improves reliability and decreases product cost. The FPGA architecture inherently requires at least one external device (the configuration PROM). On the other hand, the incremental cost may be small so the merits of the application must be evaluated on a case-by-case basis. If the application is in a system that has extra system memory, the extra storage requirements may be trivial.

One of the interesting features of the volatile nature of the FPGA is that it allows the most flexible reconfiguration of any PLD. Even when using EEPROM-based PLDs, the devices cannot generally be programmed in-circuit. (The ispGAL 16Z8 from Lattice is an exception because it is designed to remain in the circuit during reprogramming.) Reprogramming requires physically removing the device, reprogramming, and then physically inserting the device back into the board. The RAM-based architecture of the FPGA, on the other hand, means that the device can be left in the circuit. Reprogramming can also be done while the system is in operation.

One of the pressing issues in programmable logic design is the question of gate utilization. In some applications, the FPGA's ability to be reconfigured to meet different requirements can lead to gate utilizations that theoretically exceed 100%. Such flexibility is particularly useful in applications such as interfaces. Even standard interfaces often have subtle variations in functionality that can lead to incompatibility. The in-circuit reprogrammability of the FPGA makes it easy to identify and correct these types of problems in a cost-effective manner.

6.6 PLDs as Competition for Gate Arrays

The MAX and FPGA architectures often compete for the same type of applications with gate arrays. And, as one might expect, the architectures often compete with each other.

Which type of architecture is best for any given application is, of course, dependent upon the circumstances of the particular design. In this section we will discuss a few of the considerations that need to be made when deciding on which style of device to use.

As noted in the FPGA discussion, the RAM nature of the FPGA allows for in-circuit flexibility that is most useful when the specifications are not only vague but are also likely to change *in the final application.* In some applications, such as remote sensors, it is necessary to make system updates via software only. Physically reaching a probe or sensor may be expensive or impossible. Since a data channel exists, however, transferring the code to update a FPGA (along with other system software) is viable.

The routing of the FPGA also must be taken into consideration. It is possible to realize designs that have sufficient logic but for which signals "cannot get there from here." Routing also affects timing, so a certain timing variability may exist between designs that are logically equivalent. Generally, the tools available for design with the FPGAs do a good job of helping one avoid timing problems; but it is still up to the designer to consider and solve these problems.

The MAX architecture, on the other hand, is 100% routable. This is an inherent feature of the PAL-like layout. Note that this does not mean that 100% of the available logic can be utilized. The comments made in Chapter 3 concerning the utilization of resources in PALs still apply to the MAX architecture. The standard layout of the MAX device also means that the timing is essentially independent of the logic layout.

In many applications, once the design is completed there is little or no need to ever modify the circuit again. In these applications, the EPROM-like cells of the MAX device are definitely more convenient to use than the volatile FPGA cells.

Both architectures are well supported with development tools. It is difficult to objectively rate the relative merits of the available suites. Like most software development environments (or programming languages), much of one's preference tends to be based on the application and personal preference for style.

Either one of the architectures will effectively compete for designs in the 1000 to 4000 gate-equivalent range. A wider variety of MAX devices are available. Smaller MAX devices are available to compete directly with the more sophisticated PLD devices such as the PAL22V10. The larger MAX devices compete in the small to medium gate array range. The FPGA architecture is limited to the small to medium gate array range applications.

The MAX architecture is a bit more sensitive to the actual register count in a design than the FPGA architecture. This is due to the architectural layout and the way the registers are located.

For either family, the rapid turnaround times of these devices will generally make them cheaper overall than conventional gate arrays in small to medium production volume products.

6.7 GALs

The basic idea of programmable logic is to reduce the number of chips that need to be maintained in stock and populated in the circuit and to simplify the design process. With these factors in mind, it is somewhat mind boggling to think of the number of PLDs available. The designer has literally hundreds of devices from which to choose.

This large number of PLDs is, of course, opposed to the goals mentioned above. One ingenious way to simplify the decision process is to make one very flexible PLD with an architecture similar to the PAL22V10. If the device were housed in a 20 pin package and some accommodations were made, such a device could in fact replace most of the common PALs available.

This is exactly what Lattice Semiconductor has done with their popular GAL series of devices. One of the most popular GALs is the 16V8. This part can replace PAL16L8s, PAL16R4s, PAL16R8s, etc. In fact, most 20 pin PALs can be replaced by the 16V8.

To encourage the use of the GAL, Lattice designed them so that they can actually be programmed with the same JEDEC file as the common PALs. Thus if one has programs for a PAL16L8 and PAL16R4, all that is necessary to realize the design is two 16V8s. Alternately, the more flexible architecture of the GAL can be exploited by programming it as a true GAL.

Another advantage of the GAL is that these devices make use of EEPROM technology. Thus the device can be programmed over and over again. These features are obviously an advantage in the development phase of a program, but they are also quite useful on the production floor. The 16V8 allows the option of stocking one type of part even when it is not exactly clear what type of PLD may be needed. Further, the EEPROM architecture makes it possible to kit these parts for potential builds or even as spares. If the situation then dictates that other parts are required, the GALs can be reprogrammed.

The GAL style of device represents an interesting dichotomy in the world of programmable logic: The number of PLDs available that can reduce the number of PLDs needed keeps growing!

6.8 Chapter Summary

To finish our discussion of advanced PLDs, it is worth remembering that a variety of devices are available to meet any design requirement, from simple combinatorial decoding to sophisticated error correction and detection circuitry. Choosing the "correct" device for a given application is a process of trying to balance the requirements of the design against the capabilities of the device. Such considerations as the availability of development hardware and software, the relative cost of the individual components, target reliability numbers, and the projected life cycle of the product must be factored into the equation. Two inexpensive PLDs may be much better from a materials cost point of view than one sophisticated PLD. Or if design requirements are tight, the lack of flexibility resulting from choosing a PLD that is too simple might well cost more in design and development time than is saved in material costs.

Sophisticated PLDs fall into two basic categories: gate array style architectures and specialized device architectures.

- Gate array architectures are optimized for replacing "sea-of-gate" type circuitry with a single device.
- Specialized devices are normally oriented toward timing applications. Dynamic RAM controllers, sequencers, waveform generators, and graphics are typical applications.
- Both volatile (RAM-based) and nonvolatile (fuse-, EPROM-, and EEPROM-based) devices are available. All RAM-based devices are

reprogrammable in-circuit. Only a few EEPROM-based, nonvolatile devices are reprogrammable in the circuit.

- Typical gate complexities range in the 1000 to 4000 gate equivalent range. Densities may reach as high as 10,000 for some of the most advanced devices, however. Even larger numbers are projected for the mid-1990s time frame.
- Development software selection and availability is nearly as important as device architecture. A wide range of options and capabilities exists.

7

General PLD Design Issues

The material in this chapter covers a wide range of miscellaneous topics. Many of the discussions are applicable to electronic design engineering in general, but will be discussed with an emphasis on how they relate to designs using programmable logic.

7.1 Philosophy of Programmable Device Design

The purpose of this section is to provide a qualitative discussion of the philosophy of designing with programmable logic. In Chapter 13 some of the quantitative aspects of choosing programmable devices will be discussed.

The major decision that the design engineer faces in using programmable logic is the partitioning between the hardware and software. Ultimately this is an issue of cost, generally measured in dollars.

Some factors must be balanced when making these decisions. First, as a general rule, the major cost of most systems will reside in the software development. This is primarily true for microcoded or object-coded systems but also applies to most designs making use of sophisticated PLDs such as those discussed in Chapter 6.

The second factor in the decision is the cost of building the software into the system. ''Numbers are cheap'' is a phrase often heard with respect to this cost. What it means is that once the software is written, reproducing it is a minor expense.

These two factors must be balanced against hardware cost. Initially, designing hardware is generally cheaper than designing software. Hardware is more direct, more intuitively obvious, and thus easier to validate and produce.

The counter to this is that the hardware is going to drive the material costs of the product. Material costs are reduced as the number of units produced goes up, but the benefit is nowhere near as dramatic as with software. The first set of PROMs and PLDs in a prototype may well cost the design sponsor $100,000 or more. The second set of devices made from the originals will probably cost in the tens or hundreds of dollars.

These factors all form opposing vectors that settle out to an optimum solution. Exactly where this solution is depends on the application, the skills of the design team, and available resources.

As a general rule, the decision can be simplified by a single guiding principal: "If it *can* be done in software, *do it* in software." The reason for this is partly because of the cost advantage of software, but the flexibility afforded is also a major advantage. Until a product fully matures changes are inevitable. In this day and age, a product may well be obsolete long before it ever matures. This means that a product may well be changing up to the day it is replaced.

Managing these changes is a major advantage of using programmable techniques. Changing the software to correct a bug can generally be accomplished by installing a new PROM or PLD. A hardware change means expensive reworking of the printed circuit boards. Even if cuts and jumpers to the board are required, having a few extra PLD pins available can be extremely valuable. Often the extra pins can be drafted to perform some simple logic function, thus possibly eliminating the need to add another device to an already tightly populated circuit board.

The more flexible and changeable nature of software also makes it harder to document and control. Whether this is a problem or not depends upon the environment. The more complex the program, the greater the need to insure that software is adequately documented and maintained. Unfortunately, documentation is an area that is easily overlooked or is often sacrificed to make schedule and cost objectives. This can lead to expensive mistakes down the road.

Finally, one of the greatest problems people encounter while working with programmable logic is a case of an either/or attitude toward software and hardware. Modern designs are typically closely coupled networks of software and hardware. It is simplistic and outdated to think that these sophisticated systems can be partitioned into distinct camps. The performance of the hardware is directly dependent upon the instructions it is executing. The software can perform no better than the hardware will allow. In reality, like matter and energy, the software and the hardware are really just different forms of the same thing. Both have but one simple purpose: to get the job done.

In the next sections the focus will shift from the philosophical to more specific aspects of design. The first of these is designing for testability.

7.2 Design for Testability

Designing for testability is a catchall phrase that is easy to express as a philosophy but can often be difficult to implement practically. As the name implies, designing for testability means incorporating those features and techniques, from the start of the design, that make testing and validating the final product an efficient and cost-effective proposition.

Generally, the designer must consider testability from two perspectives: design validation and production testing. While it may seem that these two are closely related, in practice they are often quite different entities.

The experienced designer will try during the design process to validate the design while proceeding. This may be accomplished by simply reviewing the design mentally, having someone else review the design, simulating the design, or building a prototype of each of the major blocks of the circuit.

One of the major skills in the art of designing for testability is to know what to test and when. Too much testing shows a lack of confidence in one's design ability, slows the design process, and increases overall costs. These concerns must be balanced against the risks of committing to a wrong design approach, producing a workable but nonproducible design, or having an excessive number of bugs appear in the production product.

These mistakes are most common in a large design team environment where the overall product development effort is divided into specific groups for design, manufacturing, testing, and product support. The partitioned efforts can make it hard to insure that a change in the prototype is correctly propagated to the final product.

The easiest and most cost-effective place to correct production, maintenance, or reliability problems is often at the design stage. Unfortunately, most designers rarely get a chance to work a design all the way from concept to customer use. Therefore the people who can have the greatest effect on the overall quality of the end product often have not had a chance to fully develop the skills necessary to anticipate these future problems.

The clearest symptom of this is the development of a working but nonmanufacturable design. Consider this common scenario: Once a unit has been proven to work in the engineering lab, the designer is sent off to begin a new design. The design is turned over to manufacturing who discovers, only after committing large sums of money in printed circuit boards, components, and assembly time, that the preproduction units do not work. A frantic call is made to engineering to request support from the original designer, who, having completed the latest design, has been sent to Antarctica on a long-term consulting agreement.

After a series of hideously expensive calls to the bottom of the world, the designer is located. The designer's only comment to the production team is, "But the lab unit works!"

This scenario is somewhat expected in analog design, but is surprisingly encountered often in digital designs. The problem generally stems from one of two sources. Either something is left out during the design transfer process or the parametrics of the lab unit are such that the design is overly sensitive to some particular parameter or combination of parameters.

A common way for information transfer to break down in systems using programmable logic is caused by a failure to insure that the latest source revisions are documented. It is quite easy, particularly during the final test and checkout phase, to make minor tweaks to a PLD or microcode device. If these last minute minor changes are not accurately reflected in the release package, the changes can be lost. This produces the common refrain of, "I thought we fixed that problem! How could it suddenly reappear?"

This is particularly likely to happen when there are a large number of programmable devices in the system and the code is tightly controlled. Making sure that the revised engineering files make their way back to the controlled release can be a tedious process. Proper documentation procedures and an emphasis on good software techniques are the only sure ways of minimizing these problems.

An oversensitivity to a particular parameter in a design is a more general engineering problem. Generally, any semiconductor device will have a better than specified range of performance. For example, if a device is rated to operate at 10 MHz, it will almost always work at 12 MHz or even 15 MHz. This is particularly true if the device is being operated at room temperature.

There are many reasons that the manufacturers rate the device conservatively. Among the reasons are (1) that the parametrics will vary over temperature; (2) a desire to improve lot yields; and (3) a desire to insure that the part will operate over a wide range of applications.

If our hypothetical lab unit inadvertently makes use of this extra performance range, production problems can result. The lab unit may well get a particularly fast device. The production units, getting a more Gaussian distribution, may show sporadic and intermittent failures.

The most common parameter specification that is violated is timing. As a general design practice, timing should be derated by 25% to 50%. In other words, a device that is specified for 10 MHz should never be operated above 7.5 MHz and preferably never above 5 MHz. When this rule cannot be applied, one must be particularly careful to insure that critical timing requirements are not violated.

This caveat is particularly important for those PLDs whose timing will vary slightly as a function of the programming. If these devices are operated at or near their maximum specifications, special care is required when modifying the programming.

Designing for testability must take all of these factors into consideration, as well as several architectural issues. Consider, for instance, the portion of a circuit shown in Figure 7-1. The block marked "Divide by 128" is a common frequency scaler. This type of circuit is commonly found in communications circuits to scale system clocks down to standard communication protocol frequencies. The most typical application would be in baud-rate generators.

For our immediate discussion the specifics of the architecture are not important. The divider could be a PLD by itself or part of one of the

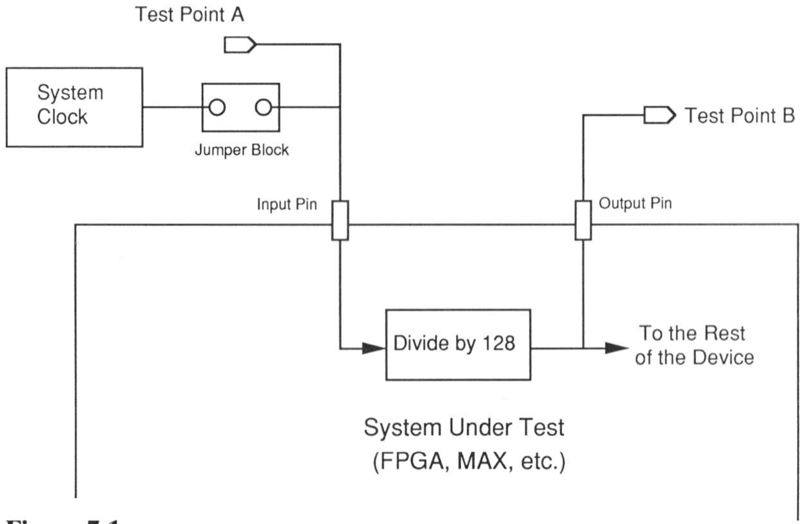

Figure 7-1
Design for Testability Example.

sophisticated programmable gate arrays we discussed in the previous chapter.

There are three basic types of testing we are interested in: *design validation, production testing,* and *fault isolation.*

In design validation, the objective is to demonstrate that the design performs as intended. This is quite straightforward if access to the output of the counter is provided, as is the case with test point B in Figure 7-1. The design validation can be done by simply putting channel A of an oscilloscope on test point A and channel B on test point B. If the period of the waveform on B is 128 times the period on A, the design of the divider is correct.

This is the intuitive approach that most engineers would take. As far as a validation test goes, this is a good way to test the design. It may or may not be a good way to perform fault isolation. For production testing, however, it has several serious flaws.

While simply looking at the waveforms seems intuitively obvious, it actually takes some fairly sophisticated observation to determine if the periods are correct. This is not beyond the capability of most *automatic test equipment* (ATE) used in production testing but it is pushing it. Worse, even if the ATE can do it, it would not efficiently make use of the test equipment.

Generally, ATE works much like the proLogic simulator previously discussed. A test vector is applied to the inputs, and the outputs are tested for the expected output. This is why the jumper is provided from the system clock in Figure 7-1. The jumper can be removed, thus allowing our ATE to apply test vectors to the input of the counter via test point A.

This brings up some interesting questions: What will be on the output when the input is toggled? How will the ATE be told what constitutes a working unit and what does not? To answer these questions, we must first consider the design of our counter, which is why it is called designing for testability.

The easiest way to build the counter is to simply chain seven flip-flops together asynchronously, as was done with the push button counter in Chapter 2.

This brings us to the first question: What happens when the circuit is powered on for testing? The flip-flops will come up in a random state. This means that as the clock vector is applied to test point A, it is indeterminate when the output will change. In fact, if the counter is not initialized at power on, it is not even possible to determine what state the output will be in. Functionally this is rarely a problem; some initial asymmetry in the output of the counter will generally be lost in the power-up sequence. By the time the system is ready to make use of the output of the counter, the output waveform will have stabilized. From a testing point of view, however, not knowing the state of the counter makes writing a test sequence very difficult.

The situation can be improved by adding a reset to each flip-flop and connecting the RESET line back to the system reset. This will insure that the output will be at a known state of zero at the start of testing and will go high after 128 clock cycles. This is something that can be programmed into the ATE.

Still, it takes 128 test vectors just to test this one function. Is there a more efficient way of tackling the problem? The answer is a qualified yes.

Consider, for example, the option of giving both presets and resets to each of the flip-flops. A typical circuit is shown in Figure 7-2. Remember, this is for *production testing;* the basic design has already been validated. Consider the following test sequence:

1. Reset the counter.
2. Test for zero on the output.
3. Preset the counter.
4. Test for a one on the output.
5. Clock the input.
6. Test for a zero on the output.

This sequence requires only three vectors. It will not, of course, prove that our counter is dividing properly. The question is, however, "what is the likelihood that our counter could have a functional problem and still pass the above sequence?" Not much.

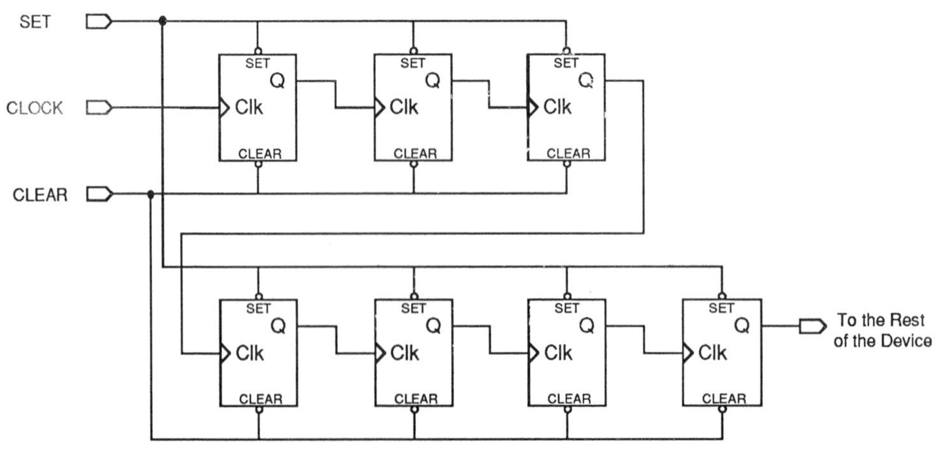

Figure 7-2
Abbreviated Functional Test Example.

Such a test is not as reliable as applying 128 vectors but it gives us good confidence with only:

$$(3/128) * 100\% = 2.4\%$$

as many cycles. Such an abbreviated functional test will typically catch 95% of all problems. The resultant savings in test time can be significant for complex products.

If independent access to each flip-flop were provided, it would be possible to preset each flip-flop individually and then test each stage of the counter. This would take a few more cycles but would give us even more confidence in the circuit.

The significance of this to programmable logic is that programmable designs are often ideal candidates for design-for-testability techniques. The PLD will often have some excess capability that is not used in the immediate application. This extra capability can be applied to producing extra test points, creating special "test modes," and adding in other special *built-in test* (BIT) features. Often little or no extra resources are required to realize these advantages.

Specific techniques include designing in MUXes to monitor critical internal nodes, adding preset and reset capabilities to individual registers, and using otherwise unused inputs to force special test outputs.

The third type of testing we mentioned was fault isolation. Whether or not abbreviated functional tests are useful for fault isolation depends on the individual circumstances. In some situations, abbreviated functional tests will not provide enough information to completely isolate a problem. Even in these situations, however, abbreviated functional tests are often useful to focus the search to a specific area.

A relatively new area of design for testability is the use of *scan registers,* often called *boundary scan registers.* The idea is similar to the points made in the earlier discussion of the counter. In boundary scan techniques, a set of registers is added to each major device in the circuit. These registers are placed in the input and output paths and are organized in a manner similar to conventional parallel-to-serial shift registers.

Normally, these registers are transparent to the operation of the device. For testing, however, it is possible to set or test each register. The idea is to allow all of the inputs or outputs around the perimeter of the test area (chip, board, etc.) to be placed under diagnostic control. This is where the technique gets the "boundary" nomenclature.

The boundary scan testing approach has been formalized by the Joint Test Action Group (JTAG). The formalized methodology provides a means of specifying industry standard techniques for this type of testing.

The data clocked out of the registers is often analyzed with a technique called *signature analysis.* Essentially, signature analysis works by computing the *circular redundancy check* (CRC) value of the serial data stream. The board is tested by applying a set of test vectors for which the CRC (the signature) is known. If the CRC is not the same as the known good signature, the board is obviously bad.

Finally, while this discussion has been oriented toward high-level issues, the basics should not be ignored. These include avoiding the use of asynchronous presets and clears for anything other than initialization, avoiding untestable internal nodes, and trying to avoid race conditions.

7.3 Metastability

Metastability is the electronic equivalent of flipping a coin and having it land on its edge. In other words, the coin toss results in neither heads nor tails. The same thing can happen when data is clocked into a flip-flop. To illustrate the problem, let us look at the circuit in Figure 7-3. This is a basic D-latch used in a variety of logic circuits.

Chapter 2 discussed the basic D-latch and its operation. To briefly review, the data is clocked into the RS portion of the latch when the clock goes high. The data is immediately visible at the output of the latch. The data is also stable. When the clock goes low, the output of the latch remains in the state it was in prior to the clock going low. This stability is achieved by the feedback provided via the cross-coupling of the two gates in the RS latch.

So far so good. The problem occurs if the data is changing at precisely the same time the clock is going low. There will be a brief time while the input to the latch is neither high nor low. This will break the stable feedback condition, and the outputs of the latch will float to a halfway point between a high and a low. This condition will in effect be a new stability point between a high and low. The condition is not truly stable. Since the cross-coupled gates form a positive feedback system, any disturbance in this balance will be magnified. Eventually the latch will assume either a 0 or a 1 condition.

The period of "sort of stable" around the halfway point is called metastability. The metastable condition can last from a few nanoseconds (in which case it is generally inconsequential) to a large fraction of a second.

Metastability is generally a result of one of two things. Either the circuit is poorly designed, allowing the data and the clock to change at the same time, or the input is asynchronous to the clock and therefore the data changes cannot be controlled with respect to the clock.

The significance of a latch going metastable is highly dependent upon the design of the circuit. The question then is "what is the probabil-

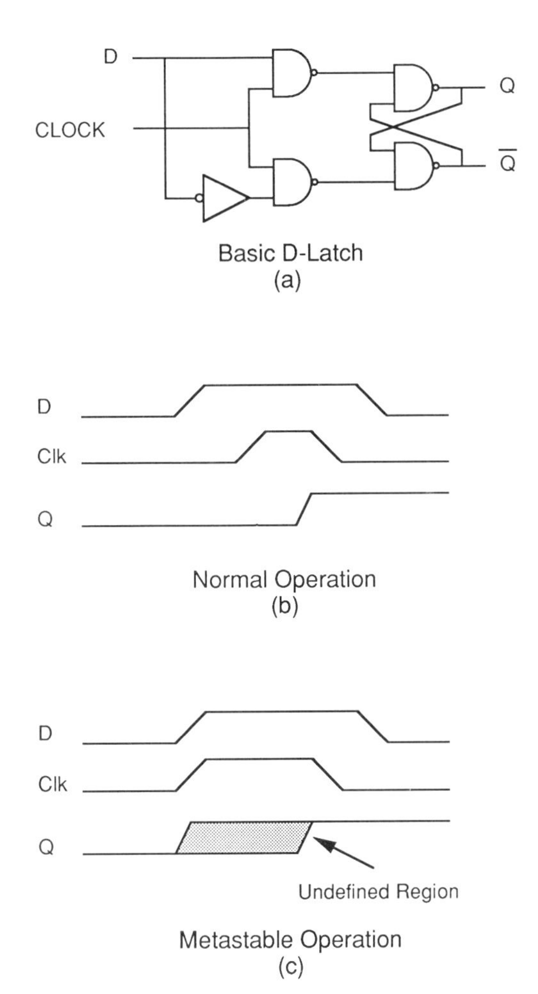

Figure 7-3
D-Latch with Normal and Metastable Operation.

ity that the circuit could become metastable, and what are the consequences if it does?'' Determining the real probability of the circuit going metastable is a formidable theoretical task that requires a great deal of knowledge of the technology in which the gates are implemented. In general, however, the chances of a conventional circuit going metastable are one in several million. This can be raised to one in several billion by good design practice and/or the use of specially designed ''metastable-hardened'' devices.

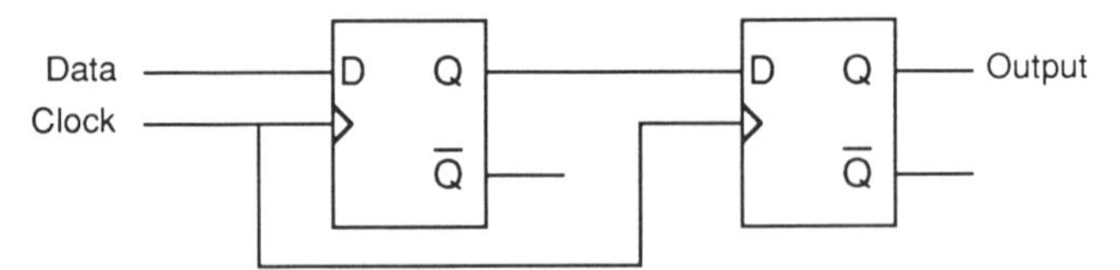

Figure 7-4
Multistage Circuit for Minimizing Metastability.

For the counter circuit shown in Chapter 2, a one in several million possibility means that there is virtually no chance that the circuit would ever experience a metastable condition.

For very critical circuits, such as life-support circuits, a one-in-a-million risk may be too high. In these cases, special care must be taken whenever an asynchronous input is presented to a state machine. For these situations, a metastable-hardened flip-flop should be specified.

Additionally, the design should be analyzed and partitioned in such a fashion that if a metastable condition does occur, the damage is limited. In most cases, metastable conditions will persist for no more than a few hundred nanoseconds. Adding an additional stage of flip-flops will therefore prevent most metastable conditions from propagating. A typical two-stage synchronizer is shown in Figure 7-4. Such an extra stage will slow down the response of the circuit considerably, but if metastable conditions are dangerous, the performance penalty may well be worth it.

This discussion has made no pretext of claiming that metastability can be completely eliminated. We have only claimed that we can reduce the likelihood of it occurring or minimize the impact if a metastable condition does occur. For any latch circuit it is possible to theorize some condition in which the circuit may become metastable.

As circuit speeds continue to increase, the significance of metastable conditions will continue to become more of a problem. Fortunately, over the last few years considerable attention has been applied to this vexing problem. New solutions and improved devices are constantly being developed and will hopefully keep pace with the increasing operating speeds.

The subject of increased operating speeds brings us to the next topic: high-speed circuit design.

7.4 High-Speed Circuit Design

PLDs are quite often used in address decoders and other high-speed applications. In order to handle these applications, the engineer working with programmable logic must have a basic understanding of the techniques necessary for handling high-speed board layouts.

Many circuit designers make the mistake of assuming that the highest frequency oscillator on the board represents the maximum design frequency. In other words, if a 6 MHz oscillator is the fastest frequency source, then the designers assume that they must design a 6 MHz system.

The fallacy in this is that 6 MHz is the *fundamental* frequency not the highest frequency. As the classic Fourier series demonstrates, the fundamental frequency is by definition a sine wave. Digital systems, however, switch with square waves (or as close to a square wave as we can get, anyway). This means that a square wave will contain many higher order harmonics. To correctly propagate a square wave, the medium must effectively deal with these higher frequencies.

This situation is illustrated in Figure 7-5. The drawing shows a square wave being propagated through a bandwidth-limited medium. In practice, the parasitic capacitances and resistances of the components and the connections on the printed circuit board act to form a low-pass filter. The effects of this limiting of the bandwidth are shown at the output for several multiples of the input frequency.

Figure 7-5a shows the effects for a bandwidth slightly greater than the input frequency. In this case, all of the harmonics are eliminated, and the output is simply a sine wave at the fundamental frequency.

The second case is shown in Figure 7-5b. A bandwidth slightly greater than three times the input frequency results in the fundamental

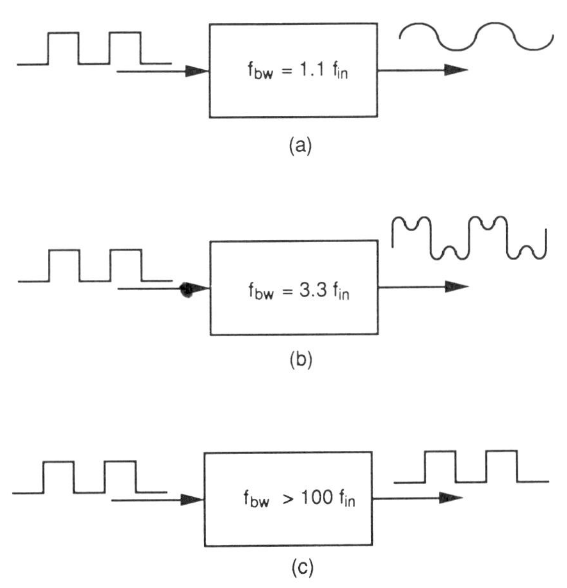

Figure 7-5
Effects of Bandwidth Limiting on Square Waves.

and the third harmonic (square waves only contain odd harmonics) propagating through the media. The result is something that is much closer to a square wave but still contains a large amount of ripple.

Finally, Figure 7-5c shows a transmission media with a very wide bandwidth. The square wave is propagated with almost no distortion.

The moral of this is that the actual frequencies in a circuit will be many times the fundamental frequency. A digital circuit using a 5 MHz square wave should be designed with the same care one would apply to a 35 MHz analog circuit.

Another important aspect of high-speed design is the rise time of the signal. As the rise time of the signal approaches the propagation delay of the signal path, special care is required. Within this timing region, the signal path will be a transmission line with a characteristic impedance. If the characteristic impedance of the source, line, and load are not matched, a signal will reflect back and forth along the transmission path.

PLDs with unbuffered latches are the most susceptible to problems from these reflections. The reflected signal can actually cause the driving flip-flop to change state. Fortunately, modern PLDs have buffers on the output latches that minimize this problem.

The reflections are still a problem to the devices being driven, however. High-speed logic will see the wave reflections as a series of pulses rather than as a single transition. This may cause transitions to incorrect states and other circuit malfunctions.

There are two classic solutions to this problem that are shown in Figure 7-6. Figure 7-6a shows the use of a series resistor at the source. The sum of the resistor and the output impedance of the driver should equal Z_0, the characteristic impedance of the transmission line.

Since the resistor is in series with the driver, this technique is called series termination, even though the resistor is at the source. An important and often misunderstood aspect of series termination is that it only works reliably with a single receiver at the *end of the line*. This can be understood by remembering that the signal is propagating down the transmission as a *wave front*.

This wave front will see the series resistor and Z_0 as a voltage divider. Thus as the wave propagates down the line, the potential of the propagating wave will be at one-half the driving voltage.

As long as there is a single receiver at the end of the line, this is not a problem. When the wave hits the end of the line, it will be reflected back immediately and the incident and reflected wave will sum to the driving voltage.

If a receiver is physically located halfway down the transmission line, however, it could easily experience problems. This can be seen by imagining the propagation of the wave and the resulting effects:

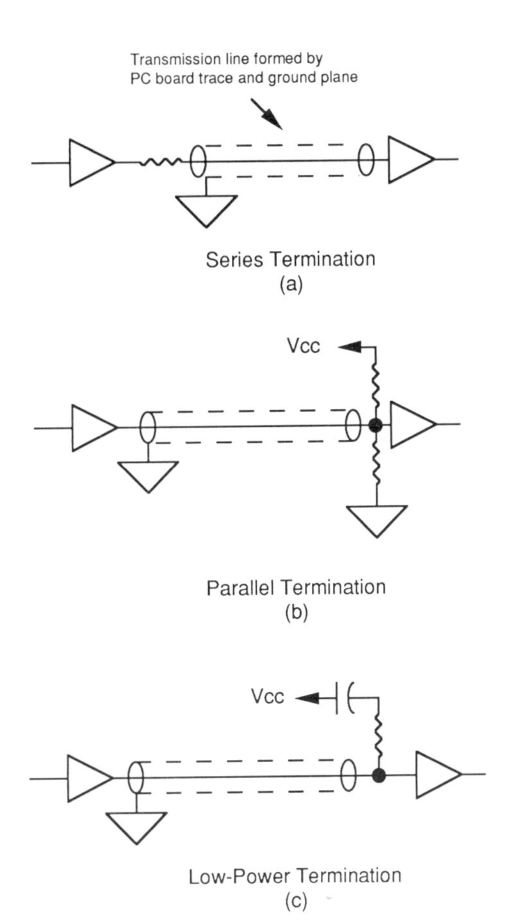

Figure 7-6
Transmission Line Termination Techniques.

- Before the wave arrives, the receiver will see a ground condition on its input.
- When the wave arrives, the receiver's input will be raised to $V_{drive}/2$. This is a noise region for most circuits and allows false triggering and possibly even oscillation. This condition will last until the wave reaches the end of the line, is reflected, and propagates back to the midway receiver.
- When the reflected wave reaches the input or the midway receiver, the voltage will rise to V_{drive}. This is a valid logic condition, but the damage has already been done.
- The wave will then travel back to the source where it will dissipate at the V_{drive} potential.

The advantage of series termination is that it uses virtually no extra power and it requires only one component (the resistor). For PLD applications, such as decoders, there will generally only be one device being driven. In these situations, series termination is generally the optimum solution.

For those cases where many devices are being driven, the termination technique of choice is shown in Figure 7-6b. The idea is that the line is biased to some optimum switching point, typically 3 V. From an AC perspective, the resistors are in parallel since the decoupling capacitors of the power supply will act as AC shorts. The resistors are chosen such that their paralleled value is equal to Z_0. Typical values for the resistors are 220 Ω for the top resistor and 330 Ω for the bottom resistor.

This technique works well and since the termination is at the terminus of the line, receivers can be placed at various points along the path. The disadvantage of this technique is that it dissipates a great deal of power.

A solution that is rapidly gaining in popularity is the use of a single resistor—equal to Z_0 or some small multiple of Z_0—with the DC portion blocked by a small capacitor. This technique is shown in Figure 7-6c. Typical values are 150 to 450 Ω for the resistor and 0.01 to 0.1 μF for the capacitor. The optimum values will depend upon the board, the technology, and other factors. The idea is to make R as large as possible, and C as small as possible, in order to minimize DC dissipation.

The physical layout of the board must also be considered. Every engineer has had the "keep routing direct and as short as possible" rule drummed into them. In practice, this rule must be tempered to accommodate high-speed layouts. High-speed lines *should* be kept as short as possible and spaced as far apart as possible. However it is often more important to insure that no "stubs" or "branches" are present on the transmission line.

This situation is illustrated in Figure 7-7. Figure 7-7a shows a driver (located at point A) driving two receivers (located at points B and C). The traces are direct, but if the lengths are moderately long (generally longer than 7 to 10 in.) problems can result. These problems stem from two sources. The first problem is the right angles of the traces. The U.S. government has gone to great expense to eliminate right angles on Stealth Technology aircraft. The reason for this is that right-angle conductors reflect energy. This lesson should not be lost on the PC board designer. The right angles of the traces will reflect part of the wave back down the transmission line, resulting in an increased noise level. The second source of trouble is that there are two receivers at the end of the branches (Figure 7-7a). Since the propagating wave will hit the inputs of the receivers at different times, the reflecting waves will be out of phase. This phase difference will also increase the noise on the line.

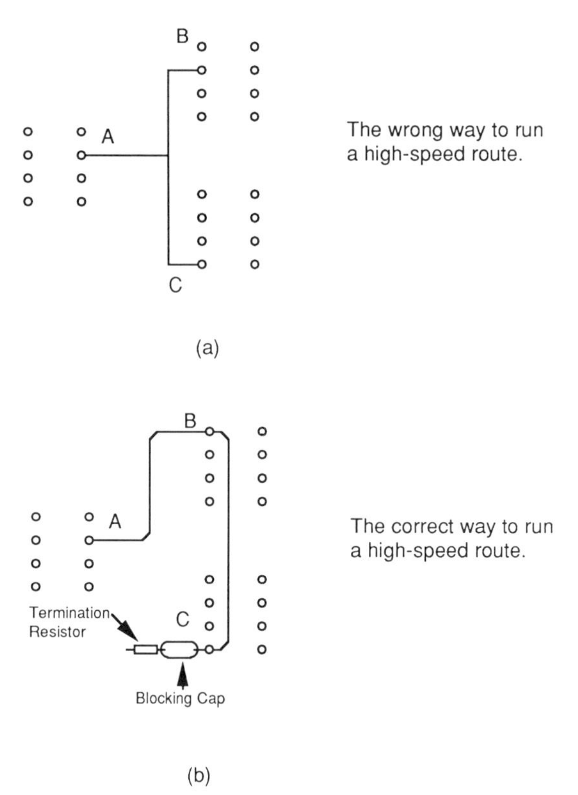

Figure 7-7
High-Speed Routing Techniques.

The solutions to these problems are shown in Figure 7-7b. The right-angles have been eliminated. The routing has also been run so that, electrically speaking, it is a straight line from the source to the terminating network. With this arrangement it makes no difference where the receivers are located along the line.

There are other aspects of high-speed design as well. One is the importance of insuring that the delays are balanced for switching paths. Another is the use of *clock trees* to insure that a clean, strong clock is delivered to all parts of the circuit at the same time.

Fortunately, programmable logic often eliminates many of the problems associated with high-speed design. The orthogonal nature of devices such as PALs makes it easier to insure that propagation delays will be uniform for various input combinations.

At the time of writing, PALs are available with propagation delays as short as 2 nsec. As these devices and other high-speed PLDs become more popular, the need for a good understanding of high-speed circuit design techniques will become an integral part of programmable circuit design.

7.5 Security

The programming of the PLD is what adds value to the part. As previously noted at several points in this text, this value-added software is somewhat dichotomous in nature. On the one hand, the programming is usually the most expensive element in the design; while on the other hand, it is the ease and low cost of changing the software that makes programmable circuitry so attractive.

This dichotomy makes for some interesting problems. Consider, for example, popular consumer goods such as electronic games. The retail price of the game is likely to be well under $30.00. The cost of the microcontroller running such a game will typically only be a few dollars. The value of the programming, however, may easily exceed $100,000.00.

This means that an unscrupulous competitor could buy such a game, copy the microcontroller code, and produce a "clone" for a fraction of the cost that the original developer incurred.

There are laws preventing such unethical things, but the enforcement is lax and the cost of prosecution must often be born by the victim. So, to prevent the unauthorized copying of programmed devices, most manufacturers have incorporated a *security fuse* into their products.

To understand the use of the security fuse, it must be remembered that most PLDs are based on PROM-like architectures. And like PROMs, the code stored in the PLD can be read back. This ability to read back the code is useful for insuring that the PLD was programmed correctly.

Unfortunately, this also makes the unauthorized duplication of a PLD quite straightforward. The security fuse is designed to prevent such an unauthorized duplication. The exact mechanism depends upon the device. For example, in some PLDs the security fuse simply disables the read back circuitry. Once the security fuse has been blown, the read back circuitry no longer allows access to the programmed array.

In other PLDs, such as some microcontrollers, the security fuse may erase on-board EEPROM and RAM before allowing access to the internal memory. If critical parameters such as address tables and key constants are stored in the EEPROM, then allowing access to the actual ROM program store is harmless.

In any case, the purpose of the security fuse is the same: to make it more expensive to make an unauthorized copy of the part than it is to produce the original code. Notice that this is not the same thing as saying that the security fuse inherently stops copying. Like any locking mechanism, security fuses can be defeated. But like conventional locks, they will often keep out the bad guys.

Finally, a comparison with the conventional "copy-protected" software must be made. The term "copy-protected", when applied to a conventional computer, is somewhat of an oxymoron. By definition, any program that *will* run on a standard computer *can* be copied. Thus, virtually all schemes to limit the unlicensed use of software have the end effect of inconveniencing the legitimate user while doing little or nothing to stop the unlicensed user.

This is not the case with the security fuse. The legitimate owner of an automobile will not be inconvenienced in the slightest if they cannot copy the code used by their engine control computer. Securing this code *can* protect a substantial investment by the producer of the engine control computer, however.

7.6 Chapter Summary

- When using programmable logic techniques, a major decision is the partitioning between the hardware and software.
- Software is generally more expensive to write but cheaper to include in the product.
- Hardware is generally cheaper to design but is more expensive to include in the product.
- In general, if the functionality can be placed in software, it should be placed in software.
- Programmable devices provide an efficient way of managing change.
- It is more efficient to design testability in from the very beginning of the design process.
- Where possible, include abbreviated functional test that provides high coverage with limited testing.
- Any registered element can be susceptible to metastability.
- When dealing with asynchronous signals, metastability cannot be eliminated. It can be reduced to any practical level, however.
- Programmable logic design often requires high-speed circuit design techniques. Keep traces short but avoid sharp bends, vias, and branches.
- Any time traces are longer than 6 to 12 in. or for very high-speed designs, consider the need for terminating networks on fast signal lines.
- If the programming in a PLD may be susceptible to pirating, use the built-in security of the PLD to limit the risk.

8

Variations on the Theme

In this chapter, we will look at some of the more interesting and sophisticated applications of programmable state machines to *logic* design. The role of programmable logic in the design and development of the control sections of computers has a long and rich history. Nowhere is this more true than in the reduced instruction set computer (RISC, pronounced "risk") versus the complex instruction set theory (CISC, pronounced "sisk") debate.

The RISC versus CISC debate has been raging for years, and it is likely to continue for many more. Central to this debate is the use and nature of a special form of the state machine known as the *microsequencer*.

8.1 Microsequencers

From the previous discussion of state machines, the reader may recall that the classic state machine can be thought of as providing 2^n possible states, with each state having m possible values. The value n is simply the number of conditional inputs and m is the number of feedback lines. The concatenation of n and m form a unique address.

For example, in the case of a classic state machine implemented with a PROM and a latch, the concatenation of n and m is simply the address into the PROM. Since each output word of the PROM is addressed by a unique address, it follows that every output of the PROM can be reached by any single address formed by the concatenation of n and m. This somewhat formal sounding description is important for three reasons:

1. The sum of $n + m$ defines the size of the PROM needed to implement the state machine.
2. Implicit in the description is the fact that any state can be reached from any other state.
3. The number of outputs is independent of either the number of states or the number of conditional inputs.

A 16 bit counter, for example, must have an m of 4. If the counter has one conditional input, say to select up or down counting, then the value of n must be 1. By definition, the count can be changed from up to down counting between any state transition, since any state can be reached from any other state.

The ability to immediately reach any state is both a strength and a weakness of the conventional state machine. What it means in practice is that any state can be reached in only one clock cycle. In the case of many circuits, bus arbiters are an example, this can be a real advantage. On the other hand, when a sequence of events is required for every state transition, the feedback state m is often underutilized.

In practical terms, for a sequence, the next state is simply equal to the current state, m, plus 1. For example, state 1 may simply point to state 2, which simply points to state 3, etc. In such a case, the feedback terms of the state machine are just implementing a counter. For efficiencies sake it makes more sense to use a counter to generate the next state information.

A basic example of this can be demonstrated in the design of an *arbitrary waveform generator*. Such a circuit is shown in Figure 8-1. The way the circuit works is as follows:

- A frequency, equal to the magnitude of the counter times the desired output frequency, is input to the counter.
- The counter simply counts through all of its states. At the maximum count, the counter rolls over to zero and keeps counting.
- The count from the counter is used as the low-order address into the PROM.
- The high-order addresses are provided by switch inputs. These allow the selection of different waveforms.
- The output of the PROM is fed into a digital to analog converter (DAC).

The waveform coming out of the DAC can be set by programming the individual address in the PROM to the appropriate value. The advan-

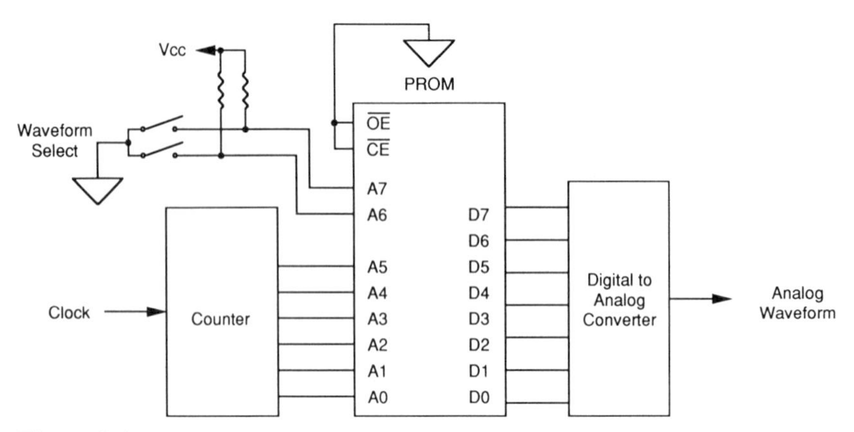

Figure 8-1
Arbitrary Waveform Generator.

tage of this circuit is that it can generate waveforms that are difficult or impossible to produce by purely analog techniques. Furthermore, such a circuit is inherently stable and essentially free from drift and distortion.

The PROM and counter combination are referred to as the sequencer portion of the circuit. Theoretically, any sequential circuit could be used as the sequencer. What is attractive about the PROM/counter combination is the simplicity of the design. The counter generates the next state, and the PROM stores the value of the output waveform for that state.

The sequencer is a flexible circuit and it need not necessarily drive a DAC. The output could drive a stepper motor, LEDs in a display, or anything else that requires a periodic sequence.

The limitation of the sequencer is that it is by definition periodic. The counter simply keeps cycling through the same states, in the same sequence, over and over. The output can be changed by changing the high-order address bits, but the sequencing will always be the same.

The PROM/latch-based state machine, on the other hand, provides flexible state transitions. Any state can be reached from any other state. The cost for this flexibility is the large number of product terms utilized, often just to provide the counter function realized in the sequencer.

The optimum solution for many designs is a combination of the two approaches: a counter for generating sequential states, but one that can be loaded like the latch of a state machine. Such a combination is called a *microsequencer*. A simplified version is shown in Figure 8-2.

There are several key points to note in Figure 8-2. First, there are three types of output from the PROM: (1) control bits (as in a classic state

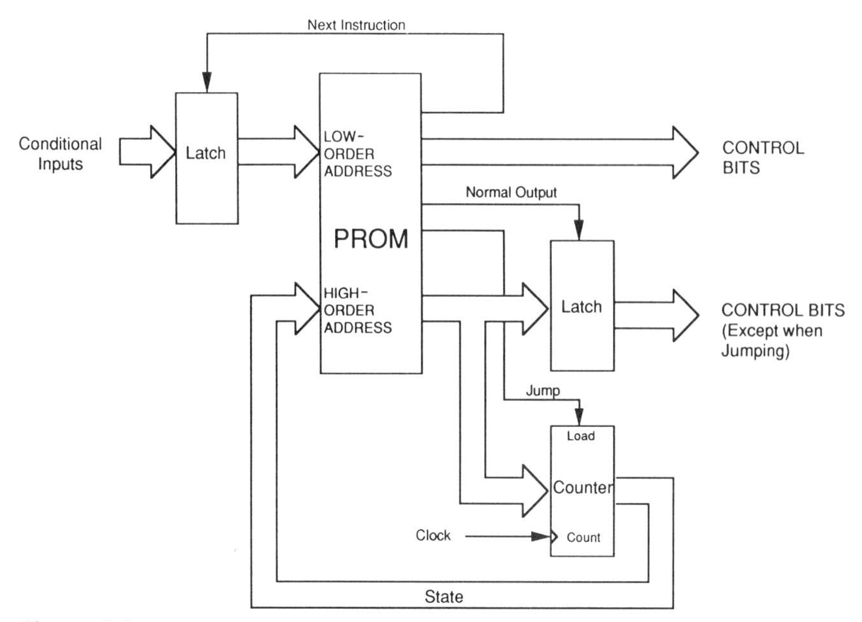

Figure 8-2
Simplified Microsequencer.

machine); (2) control/jump bits that serve a different function at different times; and (3) special function bits (NORMAL OUTPUT, JUMP, NEXT INSTRUCTION) that are used to control the microsequencer's *own* circuits.

These three special function bits are of particular interest. The JUMP line is perhaps the most interesting. A portion of the PROM's output is routed to the input of the counter. If the JUMP bit is active, the circuit acts the same as a classic state machine. If the JUMP bit is not set, however, the counter will simply increment on each clock cycle. The NORMAL OUTPUT bit, as the name implies, normally selects the control/jump bits as control bits. The data out of the PROM is normally passed on as control information to the target circuits.

This information is latched so that control information remains stable during cycles when the sequencer is jumping. The final special bit that is of interest is the NEXT INSTRUCTION bit. This bit allows the microsequencer to ignore conditional inputs until it is ready to begin a new sequence.

Microsequencers typically provide a wide range of additional functions as well. For example, microsequencers often provide a separate counter that can be used to control the number of times a loop is executed. Also common is a simple stack. The stack can be used to store addresses, making it possible to execute microsubroutines. (See Chapter 11 for more on stacks and subroutines.)

Typically, all of a microsequencer's functions are combined into one or two ICs. A typical implementation is shown in Figure 8-3. The microsequencer drives the address lines of the PROMs. The output of the PROMs are fed to special latches called *pipeline registers*. The purpose of the pipeline register is to buffer the data from the PROMs. By doing this, the microsequencer can be generating the next state information at the same time that the current state is active.

AMD's 2901 series is the most widely known of the microsequencer families. These microsequencers were specifically designed for use in the CPUs of computers. However, the 2901 series has been adapted to a wide range of applications, including digital signal processing (DSP) circuits and general purpose control tasks.

While the microsequencer approach is a powerful and flexible option, it can be seen in Figure 8-3 that a large number of ICs are usually required. In order to provide a more attractive option for systems requiring a microsequencer, Wafer Scale Integration developed their user configurable microsequencer, the SAM448. A block diagram of the SAM is shown in Figure 8-4. The SAM448 uses a PLD for the next state generation and a PROM for storing the output pattern. Both the stack and the pipeline are built in.

The SAM448 is a handy device when a microsequencer's control and flexibility are required, but the available space and cost constrain a typical discrete microsequencer/PROM implementation.

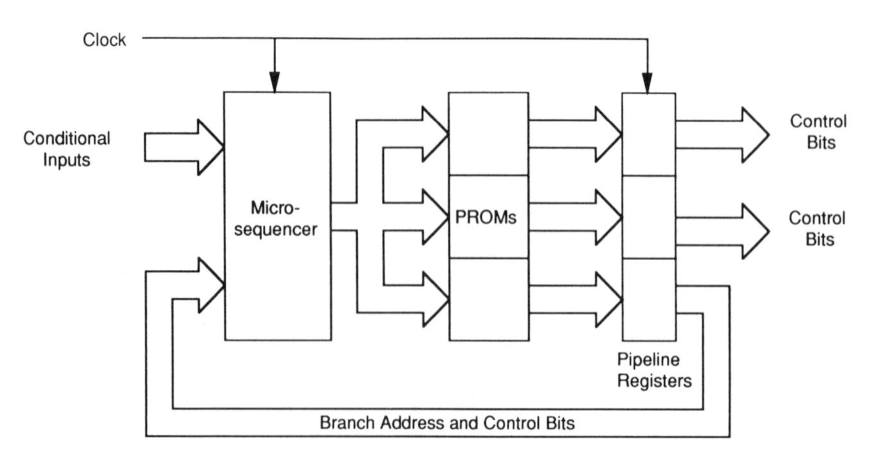

Figure 8-3
Typical Microsequencer Implementation.

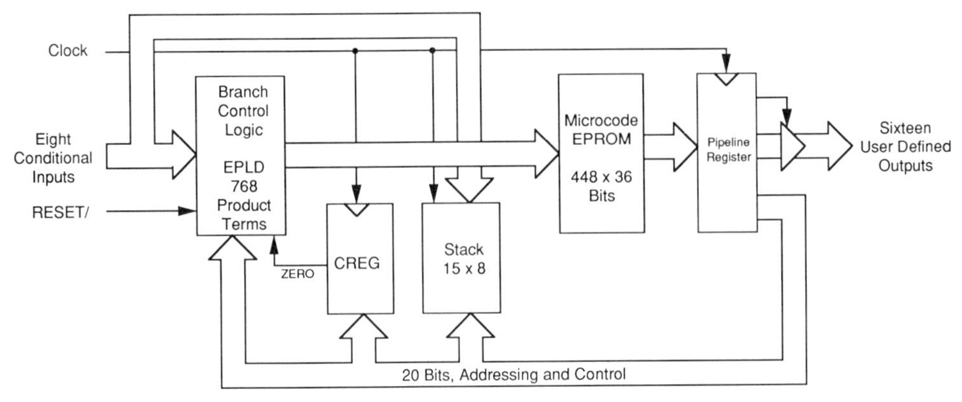

Figure 8-4
SAM448 User Configurable Microsequencer. (Adapted from the *PSD Design and Applications Handbook,* © 1990. Courtesy of Wafer Scale Integration, Inc.)

In our discussion of microsequencers, there is an explicit assumption that most steps in any given sequence will follow one after another. For many logic designs, particularly those based around processing units, this is true. A good example is the use of microsequencers in the design of a general purpose CPU. A simplified block diagram is shown in Figure 8-5.

Figure 8-5 calls out the major blocks of the CPU portion of a computer. The working registers are those normally visible (i.e., directly available) to the programmer. The address register is a sophisticated register that performs two functions. First, the address register contains a counter for keeping track of where the next instruction is in memory. Second, the output of the address register can be loaded directly, thus providing a means of addressing operands or jumping directly to different instructions. The address register can be loaded from either the data bus or from at least some of the working registers. The part that immediately interests us is the control section.

Before we discuss the control section and the important role programmable logic plays in it, we need to provide some background. In a general purpose digital computer, the instructions are stored as a series of numerical values. Each instruction is assigned a unique numerical value. For an example, we will assume a very simple set of instructions:

LDAA addr Loads accumulator A with the value stored at *addr*.
STAA addr Stores the value in accumulator A at *addr*.

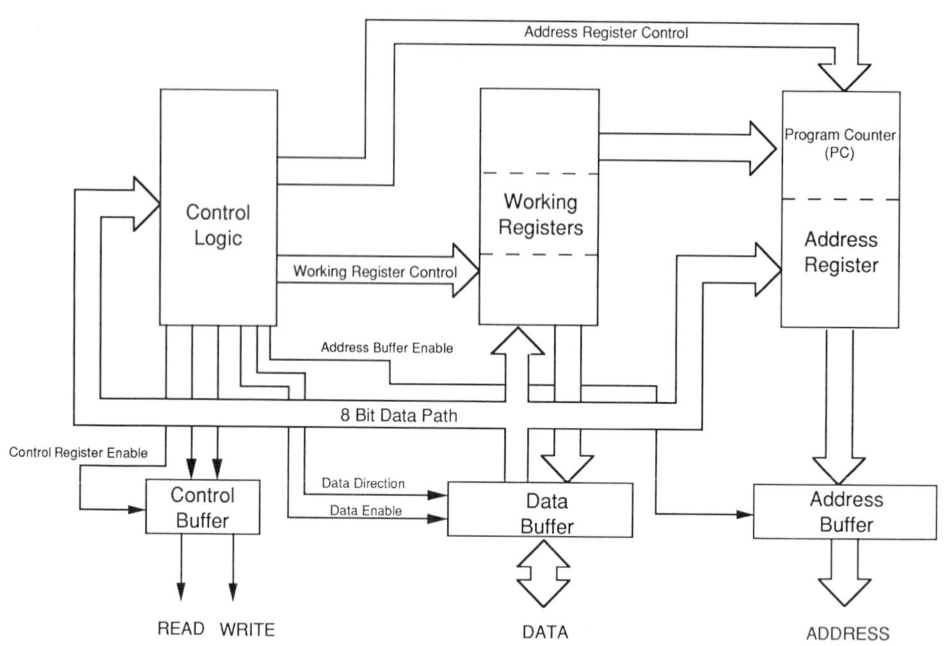

Figure 8-5
Simplified CPU Block Diagram.

The names LDAA and STAA are simply handy ways for us to remember the instructions. For this reason they are called *mnemonics*. In practice it is far more efficient in terms of memory space to simply assign a number to the instructions. For simplicity, we will assume that the machine has an eight bit wide register set, including the address register. For LDAA we will assign the hex value 01 and for STAA the hex value will be 02.

Let us assume we have a simple program. The purpose of the program is to copy a value stored in location 20_H to location 30_H. (All values in the following examples are in hex.) Assuming our program starts at address 00, it would look like this:

ADDRESS	OBJECT CODE	SOURCE CODE
00	01 20	LDAA 20
02	02 30	STAA 30

Note that only the portion called "OBJECT CODE" is actually stored in the computer's memory. The column labeled "SOURCE CODE" is simply the information from which the object code was assembled.

This is not a terribly interesting program, but we are not interested in the program itself. Our interest is in examining how the CPU's resources (registers, memory, and buffers) are controlled. The first thing is to look, step-by-step, at the sequence necessary to execute each instruction. Each step in the execution of an instruction is called a *microinstruction*.

Figure 8-6 is an example of the *control loop* used by the control section. It lists, step-by-step, the events that must be accomplished to fetch and execute an instruction. Note that the control section gets a value (which is assumed to be an instruction) from memory. The value fetched determines which instructions will be executed. If the value does not match any instruction, an *error trap* is invoked. This trap is really just another microinstruction that tells the system that an error has occurred.

The microinstructions for the LDAA instruction are shown in Figure 8-7. The microinstructions for the STAA are similar with the appropriate allowances made for buffer direction and the sense of read and write lines.

Up to this point we have not discussed how the block labeled "Control Logic" in Figure 8-5 is realized. One possibility is to simply start designing a network of flip-flops and gates that will perform the proper sequence of turning buffers on and off, setting and clearing control lines, etc. This is precisely how the original digital computers were designed. The technique is still used for relatively simple CPUs such as those found in some microprocessors and most microcontrollers.

There are some real problems with this approach, however. First, any such design is inherently complicated. A large number of bits need to be switched on and off at widely varying times. This makes a structured approach almost impossible. The lack of structure makes it hard to control the design and evaluation process. The resulting design, once tested, is virtually impossible to modify. Since the design process is hard to control and the resulting network is so interwoven, it is difficult to prove that any change will not have some other unintended impact.

These problems are not as bad today as they were for the original designers. CAD systems with sophisticated simulators now make designing and testing such a network a far more manageable process. But for the earlier developers who were limited to pencil and paper for the design process, the problem soon became unmanageable.

The programmable state machine and ultimately the microsequencer were developed to deal with these problems. Each state could efficiently implement each microinstruction. The opcode was simply connected to the conditional inputs of the state machine for decoding. Changes became much easier since control could be altered by simply changing a PROM. Almost as importantly, a change could be made in one control sequence

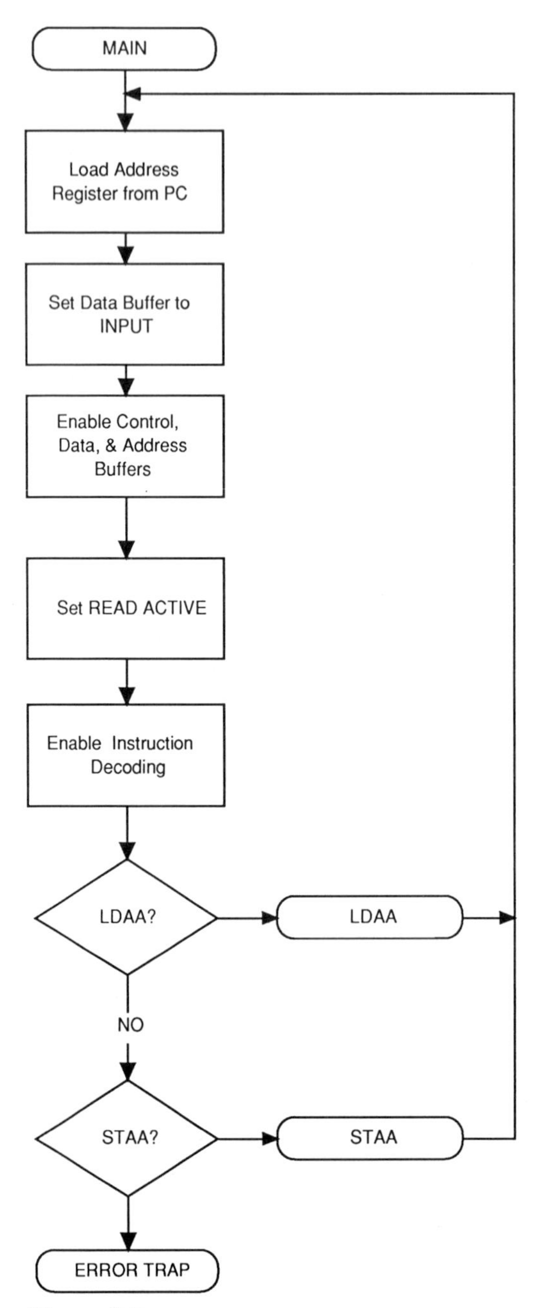

Figure 8-6
Microcoded Main Control Loop.

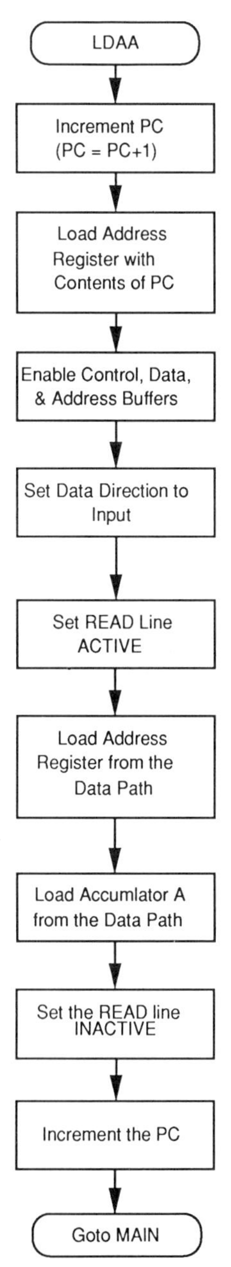

Figure 8-7
LDAA Microcoded Routine.

with great confidence that it would not impact any other control sequence.

Applying programmable state machine design techniques to the control section of CPUs was a great success. The software techniques used to control the programs running on the computers could be used to control the configuration of the CPU!

8.2 RISC versus CISC

This is a book about programmable logic, not computer architecture, so we will not follow the path into computer design any further. The main thing to realize is that at the core of virtually every sophisticated conventional CPU is a state machine. This state machine has a lot of supporting circuitry, thus giving it the name microsequencer. Since the range and sophistication of a computer's instruction set put together this way is virtually unlimited, these machines are called complex instruction set computers (CISC).

Now that we have covered the significant contributions that have been made by microcoded design techniques, we will give equal time to the opposing candidate, the RISC philosophy of design.

Reduced instruction set computers (RISC) are an outgrowth of research performed at IBM. Careful study of the type of instructions and the frequency of use of various instructions in computers led to some interesting findings. It was discovered that most of the instructions in a "typical" computer were rarely used. Many computers had complex instructions that were used to set indexing, control memory management, and perform other "house keeping" functions. Microcoding made it so easy to include these instructions that the list kept on growing, even though many of the instructions were rarely used.

From these studies of instruction set usage it was reasoned that a more efficient, faster computer could be built if one were to reduce the complexity of the instruction set (thus the name RISC). The complex instructions would be replaced by a series of simpler, primitive instructions. These simpler instructions could execute more rapidly and efficiently than their complex counterparts.

Admittedly, there are times when complex tasks such as indexing are required. In these cases, several RISC instructions are needed to execute the equivalent of one CISC instruction. The RISC instructions would take longer to execute than the CISC instruction, but the overall instruction throughput is improved. The reasoning is that those instructions most often executed would be faster, thus offsetting any time lost by the occasionally slower sequence required to "emulate" a complex instruction.

In many ways microcoding never really set well with digital designers. Not only is it software-oriented, but from a hardware perspective it is very inefficient. The memory architecture of PROMs and the supporting circuitry that goes with it are very "real estate" intensive. Designers had bowed to the use of microsequencers as a necessity.

As noted earlier, modern CAD tools make the design and development of complex hard-wired logic designs much more manageable. These improved CAD tools allowed designers to realize the simpler instructions set of RISC machines. Nothing is free, however. In the process of taking more direct control of the control logic, the instructions became wider. These wider words are why few RISC machines have less than 32 bit wide words.

Many other concepts are often associated with RISC machines. These include very large register arrays, single-cycle execution, etc. But the core issue is microcoded control versus hard-wired control. This can all be looked at from another angle. Effectively, a CISC machine compiles code down to the object code level. The CPU then uses a microsequencer to *interpret* this object code. A RISC machine effectively skips the interpreting step and compiles down to the microcode.

At the time of this writing, RISC designs are making significant inroads in certain areas but are by no means dominating the design of computers. While many of the RISC concepts will undoubtedly continue to have significant impact on computer design, it is unlikely that microcoding or state machine design in modern computers will go away.

8.3 Writable Control Stores

The PROM is a handy and generally convenient way of storing the code for a state machine or a microcontroller. Like anything, however, there are some trade-offs. Changing the code means physically changing the PROMs, for example. Further, PROM's are not as fast as other memory devices such as RAMs. The nonvolatile nature of PROMs are necessary for storing data, but when used in state machines this means that the instructions cannot be changed dynamically. For things like the arbitrary waveform generator discussed earlier, this is a serious limitation.

One solution to these problems is to use RAM instead of PROM or ROM as the program-store elements. Architecturally the RAMs are used exactly as the PROMs. The only difference is, of course, that the RAMs must have some means of having the data loaded into them.

When RAMs are used this way, the resulting architecture is called a writable control store. Often, the RAMs will be managed by a microcontroller that is connected to the source of the microcode. The source may be inexpensive PROMs, magnetic disks, or even paper tape. When the system is booted up, or anytime the control information needs to be changed, the microsequencer is halted and the RAMs are loaded.

The use of PROMs in a writable control store may seem redundant, but there are advantages other than just improving the speed. The writable control architecture allows instructions or microroutines to be patched as needed. This is particularly useful for remote sites. The remote site may simply be a customer on the other side of the country or it may be a spacecraft in a far orbit.

Writable control stores, combined with RAM-based FPGAs such as those from Xilinx, can be used to form an essentially free-form computer. It is possible to build a system where the I/O ports, the overall architecture, and even the instruction set is completely defined in software. While such a system is merely an interesting concept at this time, the long range implications are truly mind-boggling.

8.4 Chapter Summary

Programmable circuit design has played a long and illustrious role in the development of the modern computer. This development has been synergistic. Software techniques designed for use at the systems programming level have been successfully applied to the hardware design of the CPU and even to the definition and implementation of the instruction set architecture.

These techniques are not limited to computer design. The power and flexibility of microsequencer-based designs can be applied to a wide variety of digital and even analog applications.

9

Introduction to Microcontrollers

Microcontrollers are complete computer systems on a single silicon chip. They include the central processing unit (CPU), random access memory (RAM), read only memory (ROM), and peripheral circuits necessary to implement a wide variety of programmable logic functions.

Microcontrollers are a classic case of a difference in *degree* becoming a difference in *kind*. From a theoretical perspective there is no difference between a simple four bit microcontroller and a Cray XMP. Both machines are essentially classic Von Neuman machines, making use of program stores to hold instructions, data stores (RAM) to hold data, and accumulators to manipulate the data under control of the instructions.

As Turing has proven, any problem that is solvable with a computer is solvable with a one bit processor, given enough time and memory. Therefore, processors at either end of the spectrum can theoretically be used to solve the same problems.

Obviously, however, in reality there is a great deal of difference in the kinds of problems one would handle with each type of machine. Modern general purpose computers are designed to solve a wide range of problems. The computer may be used as a lab monitor, as a word processor, or as a mathematical "number cruncher." This demand for flexibility determines many of the techniques that are used in realizing a general purpose computer.

Microcontrollers, on the other hand, are designed to solve a relatively narrow range of problems known as *embedded control applications*. These applications are characterized by several key attributes:

- Typically, embedded control applications do not change. An automotive engine controller will always be an engine controller. The actual program may change to achieve higher performance or to accommodate other changes, but the application remains the same.
- Embedded control applications are relatively small. While it is not hard to find exceptions to this statement, most embedded control applications can be coded in less than 8K words.
- Even with the above two comments in mind, the range of applications is quite broad. The same type of microcontroller may be used in a child's toy, microwave oven, darkroom timer, or many other such applications.

These characteristics have a pronounced impact on how one approaches design with a microcontroller. Before looking at a few of these considerations, however, let us look at why one would use a microcontroller in the first place.

There are two main reasons for using a microcontroller to solve a design problem:

1. The design specifications are loose enough that the flexibility awarded by a programmable device is necessary. This looseness may be caused by a desire to rapidly respond to new requirements, a poor understanding of the problem, or a lack of definitive information.
2. The problem at hand is inherently better solved by software than by hardware. This is the case, for example, with such applications as calculators or microwave oven controllers. It is possible to come up with purely hardware solutions to these problems but why bother? Programming the solution is much simpler and more cost-effective.

Microcontrollers can be used in many of the same applications examined in previous chapters. Combination locks, with their wider class of pattern detectors, are only one example. In upcoming chapters we will look at the final variation on the push button circuit and compare and contrast the relative merits of each implementation.

There are also some differences in how one approaches a microcontroller design. Most modern engineers are used to doing some programming on a general purpose computer. In computer science jargon this is known as *native* programming. The reason that it is called native programming is that the computer being used to generate the code will actually be executing the code.

Generally, the code will be generated from a high-level language, typically BASIC, Pascal, FORTRAN, or C. Particularly in the case of the first three languages, the user will be "protected" from the hardware details of the machine. Type a key on the keyboard and the character automatically appears on the screen. Use a print statement and the information comes out on the printer. This allows the programmer to focus attention on "higher level" activities such as user interfaces and algorithm development.

The microcontroller, on the other hand, is not generally designed to support program development. Its inputs and outputs are likely to be real time signals such as switch settings and sensor inputs. The outputs are typically things like control lines to motors, relays, or other actuators.

There are many implications to this. First, programming of microcontrollers is commonly done in assembly language rather than in one of the high-level languages. The programming is much more hardware-oriented, with a strong emphasis on addresses, control lines, bit states, and binary operations.

Naturally, the code development and programming of the microcontroller must be done somewhere, and this is usually on a general purpose computer such as a PC. When development software such as a microcontroller assembler is running on one machine and the code will be executed on another machine, it is called a *cross assembler*. The machine that will be executing the code is called the *target machine*.

Development software and hardware can be obtained from a wide range of resources. The microcontroller manufacturer often provides suites of development hardware and software designed for their specific product. Alternately, third-party development suites can be purchased for most popular microcontrollers. These tools are discussed in Chapter 12.

When programming for general purpose applications such as PCs, execution time of the program is often not a major concern. While faster is always better, the step-by-step timing of a general application program is not critical. Due to the real time nature of many microcontroller applications, however, timing is critical. The programmer must often spend much of the coding time insuring that critical time constraints are not violated.

The demanding nature of writing code for a microcontroller combined with the need for an intimate understanding of the hardware often alienate programmers used to working with high-level languages on larger platforms. Many engineers, on the other hand, are very hardware-oriented anyway. Many engineers who have taken the time to learn how to effectively program microcontrollers feel that they have achieved the best of both the software and hardware world. Indeed, in no other area is the line between software and hardware as blurred as it is with microcontrollers.

Microcontrollers contain a wide range of hardware functions. Timers, counters, serial channels, analog-to-digital converters, phase-locked loops, and other common functions are only just a few examples. These functions are complemented by sophisticated (in comparison to state machines or microcoding) software resources. Indexed instructions, arithmetic operations, and bit manipulation operations are all available for integrating and orchestrating the hardware resources.

All of these resources make the microcontroller the most versatile vehicle for implementing arbitrary logic functions. Any logic function that can be performed in hardware can also be performed by a microcontroller in software. The tradeoff for this flexibility is speed. Microcontrollers will typically take anywhere from 5 μsec to several hundred milliseconds to respond to an event. This is in contrast to the fractions of microseconds for state machines or nanoseconds for high-speed PLDs.

On the other hand interestingly enough, the dollar cost of using microcontrollers may be no greater than PLDs. The cost will likely be less than the cost of implementing a conventional state machine. In large quantities, microcontrollers can cost as little as a $1.00 apiece for factory-masked ROM units. One time programmable (OTP) microcontrollers may cost from $5.00 to $10.00 for the simpler units, roughly comparable with more sophisticated PLDs.

Microcontrollers are often the easiest type of programmable logic to work with for a variety of reasons. The rich assortment of resources available in a microcontroller often allow design requirements to be met in a variety of ways. The computational ability of the microcontroller is unique in programmable logic, making flexible solutions more readily achievable. Finally, the support tools for microcontrollers, both software and hardware, are generally more sophisticated than those available for the architecturally simpler PLDs or state machines.

Over the last several years an interesting but somewhat paradoxical situation has been developing with microcontrollers. As these devices have become cheaper and easier to use, their level of sophistication has been increasing considerably. Techniques once reserved for sophisticated mainframe computers are now being employed for even the simplest of microcontrollers. These include real time interrupts, process driven control, multiuser and multiprocessing software, and other sophisticated software and hardware techniques.

The next two chapters discuss these subjects in greater detail. Chapter 10 covers the hardware architecture of common microcontrollers. Chapter 11 looks at the software architectures typically used with modern microcontrollers.

10 Hardware Architecture of Microcontrollers

As was noted in Chapter 9, the term *microcontroller* defines a single computer on a chip. Included on the chip are all the major blocks normally associated with a computer: the CPU, memory, and I/O. Generally, the chip will also contain special function hardware. Typical examples of special functions include timers, counters, and analog to digital converters (ADCs). Digital to analog converters (DACs) are rare but not entirely unheard of on microcontrollers. A block diagram of a generic microcontroller is shown in Figure 10-1.

Microcontrollers typically fall into one of three classes: 4 bit units, 8 bit units, and 16 bit units. Like the common 8 bit microprocessors, 8 bit microcontrollers generally have some capability to handle 16 bit values.

Occasionally, other sized microcontrollers are encountered. One bit and 32 bit sizes are found in special applications. These are rare, and we do not specifically discuss these unusual units in this text. However, most of what is in this chapter will apply to any microcontroller, regardless of the size.

10.1 Basic Features

The basic features of the microcontroller [the CPU, the program store, the data store (RAM), and parallel I/O] form the basis for our initial discussion of the hardware architecture.

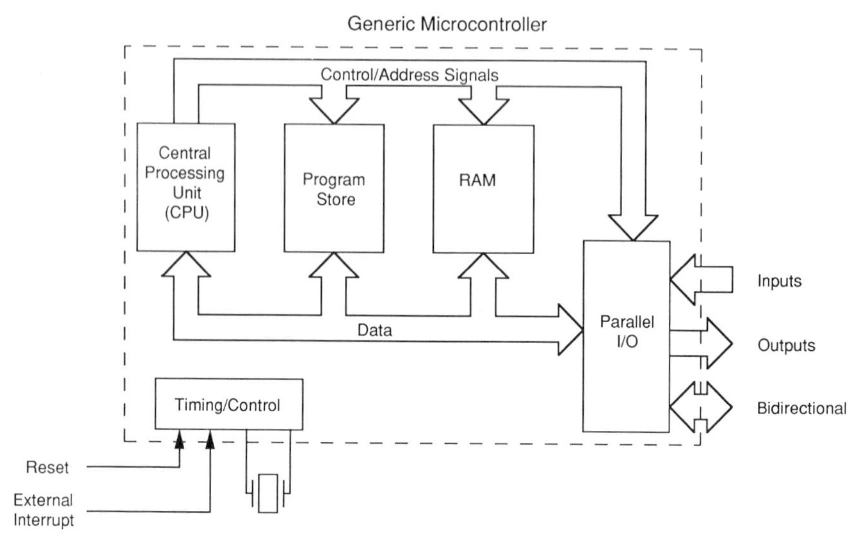

Figure 10-1
Basic Architecture of a Microcontroller.

The CPU performs all of the actual manipulation of the data. The performance capabilities of the CPU in doing this manipulation vary widely, generally as a function of the width of the instruction. Four bit processors often have simple instruction sets, while 32 bit CPUs may have quite sophisticated instruction sets providing a wide range of data manipulation capability.

In general, the instructions executed by the CPU are similar to those executed by popular microprocessors. The register load, bit comparison, and simpler arithmetic operations are basically the same, though the exact form will vary. Often the CPU lacks the sophisticated indexing registers that are available with microprocessors. This results in fewer indexing options, restricted address ranges, and a smaller variety of operators than are often found in microprocessors.

Generally the CPU's instruction set will be richer in bit manipulation operations than a general purpose microprocessor. These bit-oriented functions are of course more common in microcontroller applications. Bit test-and-set, test-and-clear, masking, and other operations are typically well supported.

The MC6801 is a popular medium-performance microcontroller found in many automotive applications. As an example of a typical microcontroller, the register set of Motorola's MC6801 is shown in Figure 10-2. There are two 8 bit accumulators, ACC A and ACC B. These two accumulators can be combined to form a single 16 bit accumulator, ACC D. A

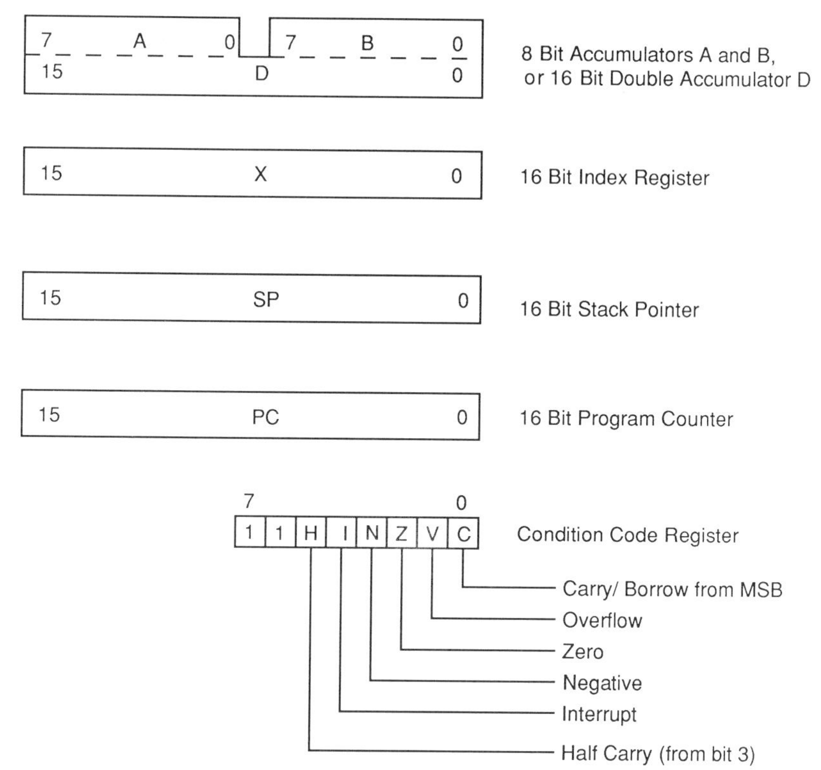

Figure 10-2
Register Set of the MC6801.

16 bit index register is available to simplify accessing data structures such as tables and strings. The stack pointer is also 16 bits and can be used to place the stack anywhere in the memory space. The flag register holds the status of the various machine states such as the interrupt enable bit and carry results.

Finally, the program counter is a 16 bit register that is used to step through the program. A JUMP instruction allows the program to unconditionally transfer control anywhere in the 64K address space. Conditional control is transferred using branch instructions that allow -128 to $+127$ range relative transfers.

The RAM in a microcontroller is generally dedicated to handling the stack and for processing variables. Due to the dedicated nature of microcontroller applications, there is seldom any need to load programs into RAM. For this reason, many microcontrollers cannot even execute code from RAM. Motorola's MC6801 series is an exception. Both the MC6801 and the high-performance MC68HC11 have 256 bytes of general purpose

RAM. As in a conventional computer, this RAM can be used for stack, code, or variables as the programmer desires.

The program store of the microcontroller will usually be one of two types: masked programmed ROM or UV erasable EPROM. Other alternatives that are gaining in popularity include one time programmable (OTP) EPROM and EEPROM. Typical densities range from 0.5K for simple 4 bit units, up to 32K for more sophisticated 16 bit or 32 bit units.

Virtually all microcontrollers will provide simple parallel I/O. This I/O generally consists of simple 74HC373 style output latches. Often, the I/O pins have input buffers that allow the state on the pins to be read in as well. Typically, certain ports will be input only, certain ports will be output only, and some ports can be used dynamically as either.

Some port configurations are selected as part of the mask programming but most are selectable at run time by the software. A typical port architecture is shown in Figure 10-3.

Most microcontrollers use some minor variation of this arrangement so it is worth discussing it a bit. The basic I/O pin is supported by two latches. One, the data latch, is used to hold data that is written to the port. The other latch is used as the "data direction register" (DDR). This term

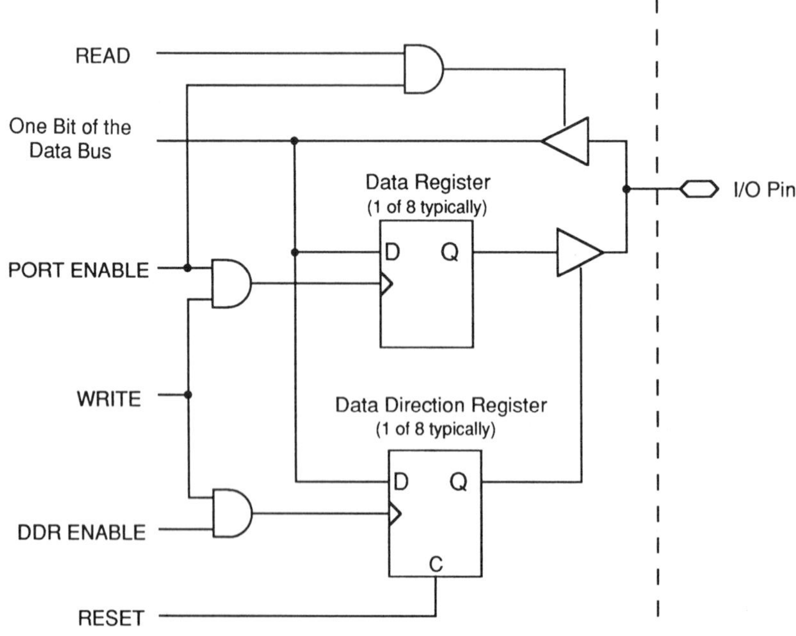

Figure 10-3
Typical I/O Architecture.

is actually a misnomer. What the DDR actually does is to enable the output of the data latch to be seen on the pin.

For example, if we want the I/O pin to be an input, we write a zero to the DDR latch via the internal data bus. This disables the three-state driver, and any external device driving the I/O pin can be read via the input buffer.

If we want the pin to be an output, we write a 1 to the DDR. This enables the output buffer and the data present in the data latch appears on the output pin. Notice that a read from the port is still valid and we will simply read back what was written to the port. This read back capability may seem redundant, but in practice it can be quite useful. The read back capability of the latch allows us to read the port value, AND or OR the bits as necessary, and then write the result back to the port. In this way the bits that were not operated on are left unaffected.

Most microcontrollers allow the bits in the DDR to be set individually. Thus a single port will often contain a mix of both input and output bits. Also notice that the DDR is cleared by the RESET of the microcontroller. This insures that all I/O pins come up as inputs.

This is often important for the controlled power on of a system. When the I/O pins are being used as control lines, external pull-up or pull-down resistors can be used to bias the lines to their inactive state. Then the microcontroller can set the appropriate control states in the data latch. Finally, the DDR can be written, and the data in the data latch will be available on the I/O pins.

10.2 Common Optional Features

In addition to the standard features on a microcontroller, a wide variety of options are available. One of the most popular features is an asynchronous serial port designed to support RS-232 type communication channels. These serial channels are very useful in master/slave configurations. We will look at a few examples shortly.

Another very popular feature is some sort of timer/counter. Generally, the input of the counter can be routed either to one of the input pins or to the system clock. This allows the same circuitry to serve as an event counter or to be used as a timer. The output of the timer can be used to provide periodic interrupts, to precisely control critical timing, or to eliminate the need for software timing loops.

Synchronous serial channels have grown in popularity over the last several years as well. One the major advantages of this type of channel is the powerful expansion capability it provides. The synchronous channel can be combined with a parallel-to-serial shift register to expand the number of inputs to the microcontroller. Or a serial-to-parallel shift register can be used to expand the number of outputs. A typical application is shown in Figure 10-4. In this example, the synchronous serial I/O channel has been used to expand the number of both inputs and outputs of the microcontroller.

The operation of the circuit is quite simple:

- The microcontroller first loads the input data by bringing the input shift registers SH/LD line low. It then sets the line high to allow the shifting to occur. In this case we have used port A, bit 0 (PA0) for this function.
- The microcontroller then writes the output data to the synchronous serial I/O data register. Writing to the microcontroller's data register automatically causes the data to be synchronously shifted out of the microcontroller and into the output shift register.
- Simultaneously, the data in the input shift register is clocked into the synchronous serial I/O data register. The input data can then be read by the microcontroller's CPU.

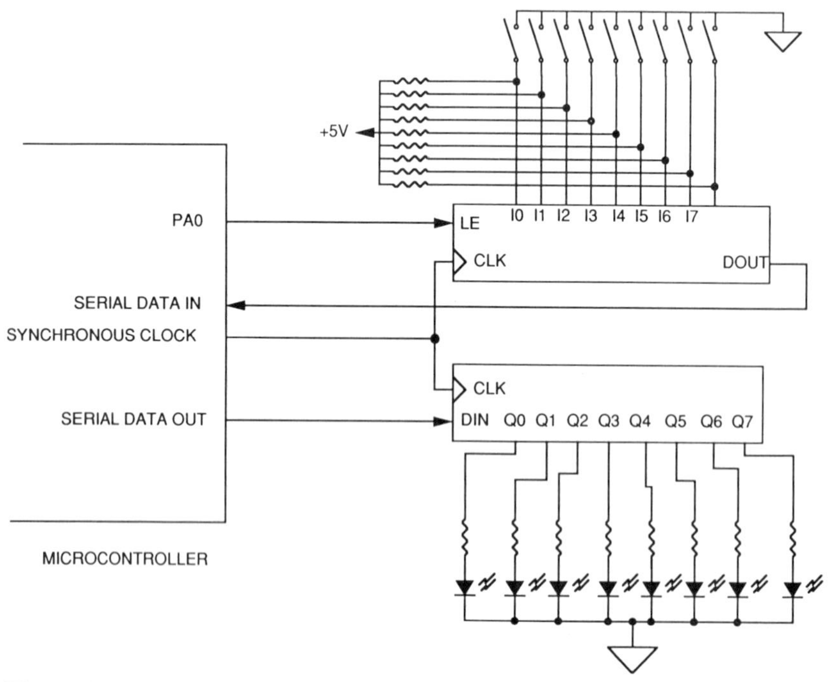

Figure 10-4
Expansion Using Shift Registers.

This scheme costs four microcontroller pins, but it allows us to monitor eight additional inputs and control eight additional outputs. Since I/O pins are often one of the scarcest resources in a microcontroller application, this is a quite useful technique.

We should note, however, that the output bits are not "glitchless." That is, as the data is shifted into them, the outputs may momentarily change. For the majority of applications this is probably not a concern. The synchronous clock frequency is typically in the range of 50 kHz to 1 MHz. If the output is driving a display such as an LED, the disturbance will not be noticeable.

More specialized peripheral chips are also readily available to expand the power of the microcontroller via the synchronous channel. Typical examples include serial RAMs, real time clocks, and serial EEPROMs.

Many options also exist for the way data is entered into the microcontroller. The real world is mostly analog. This fact has not been lost on microcontroller vendors. Most modern designs are being offered with 8 bit or 12 bit analog to digital converters (ADCs). While these converters are not as fast as discrete ADCs, many are fast enough to support low-quality voice applications.

A typical microcontroller application is shown in Figure 10-5. This is a sophisticated electronic thermometer. The actual quantity being measured could be the air temperature, an industrial process, or even the temperature of a patient.

Typically, the way such an application works is as follows: the output of the thermocouple is fed to analog signal conditioning circuitry. The analog circuitry amplifies the signal and limits the signal to a range that is within the dynamic range of the ADC. The microcontroller reads

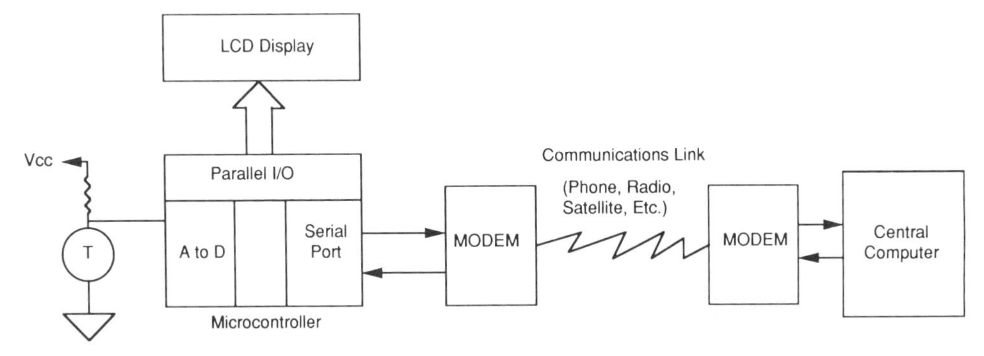

Figure 10-5
Microcontroller Application.

the ADC and obtains the raw temperature value, which is the voltage output of the signal conditioner.

The microcontroller then uses the digitized signal to compute the conventional temperature to display. The math capability of the microcontroller can be particularly handy for these computations. Or if memory is abundant, a simple lookup table can be used. The temperature information is displayed on the LCD and at the same time saved in memory. Periodically, or if the temperature exceeds predetermined bounds, the microcontroller uses the MODEM to call the central computer and report the temperature.

Such a smart thermometer can be installed anywhere that there is access to a phone line. Given the relatively easy access to cellular phones and satellite channels, this means that such a smart thermometer can be placed almost anywhere in the world.

It is worth noting that with the ever-falling cost of VLSI modems, the incremental cost of adding the remote monitoring capability is not necessarily expensive. Undoubtedly more of these types of systems will be coming. Already, many common home appliances are digital. It is only a matter of time before the TV, microwave oven, and even the bathroom scale will all be networked together with the home computer monitoring and coordinating it all.

10.3 Exotic Optional Features

In addition to the standard options, many microcontrollers are available with some fairly exotic features. Examples include phase-locked loops (PLLs), gate arrays, DACs, and other specialized circuits.

Some microcontrollers, particularly those of Japanese origin, are specifically designed to work with certain displays. The microcontroller will often have a built-in LCD or LED display controller hardware. This reduces system cost considerably when producing large volume consumer products such as calculators.

Some advanced standard cell architectures have turned the tables on off-the-shelf microcontrollers by including microcontrollers in their libraries. With this capability, the high-volume user can simply place a microcontroller on the die along with whatever other custom circuitry is necessary. This forms a powerful, best of both worlds scenario.

Even more exotic combinations are constantly being developed. Motorola, for instance, has its customer specified integrated circuit (CSIC) program. The customer presents Motorola with their specialized requirements. These are typical microcontroller cores with specialized peripheral circuitry. Based on a number of evaluation criteria, if Motorola feels there is enough interest, they will develop and produce the part at no cost to the customer in return for the rights to market the device. These types of programs stimulate the already exploding range of options available.

Now that we have looked at the general hardware architecture of microcontrollers, we will focus on some of the characteristics of the three major classes of devices. These classes are partitioned by the width (expressed in bits) of the microcontroller. As with microprocessors, the width can be somewhat confusing. The width of the accumulator, data bus, and instruction size are not necessarily the same. Generally, the width of the accumulator is used to specify the size of the device. A 4 bit microcontroller will normally have a 4 bit accumulator, but may well have a 12 bit instruction word.

10.4 Four Bit Units

The four bit family of processors are common choices for high-volume, low-cost applications. Typically, these units are mass programmed at the factory to keep the per unit cost low. Popular applications of four bit units include microwave oven controls, calculators, electronic scales, toys, and other mass market items. The great strength of four bit designs are their low cost and small size. Four bit microcontrollers can be found in 16 pin DIP packages for less than $1.00 when purchased in large quantities.

Four bit microcontrollers are relatively difficult to work with due to their limited resources. They work well enough in numeric applications since four bits is adequate for dealing with BCD encoded numbers. They are much less useful in anything requiring ASCII or other alphanumeric encoding where at least seven or eight bits are generally needed. Furthermore, the limited instruction sets combined with the general lack of amenities make program development somewhat tedious.

Due to the lack of processing power, the four bit units cannot support native code development (''native code'' is actually written and compiled on the machine that will be running the code.) So any code development must generally be done on relatively expensive development stations. These development stations can range from approximately $10,000 to over $50,000. These relatively high capital costs mean that the incremental cost savings of a four bit unit must be generated over large volumes to be cost-effective. See Chapter 12 for more information on development stations.

10.5 Eight Bit Units

Eight bit microcontrollers are the most popular of the microcontrollers. Eight bits is the smallest size that is practical to handle most data types yet still allow a high enough performance to effectively deal with most applications.

Microprocessors first gained prominence in eight bit configurations. The early i8080, i8085, Z-80, and 6800 set the stage for the personal computer revolution. While these units seemed quite powerful at the time, increasing demand for higher performance led to their replacement with more powerful 16 bit units. Today 16 bit units such as the 80286 or the 68000 are considered the minimum for even home computers. This trend will undoubtedly continue in computers, with the 32 bit 80386 and 68030 processors becoming common.

The evolutionary path for microcontrollers is quite different. Many of the embedded control applications for which microcontrollers are used do not benefit from wider word widths. The smaller memories and lower pin counts necessary on a microcontroller can often be better utilized with eight bit instructions.

Improvements in microcontrollers have tended to focus on more powerful instruction sets (flowing against the RISC tide), larger memories, and more powerful on-chip peripherals.

The performance capabilities of eight bit microcontrollers range from only slightly above those of four bit units to being on a par with some of the better microprocessors. Many eight bit microcontrollers can, in fact, be used as microprocessors. Typically, one eight bit bidirectional port is used as an eight bit multiplexed address/data port, and a second port is used as an additional eight bit address port. This capability has three distinct benefits:

1. It allows conventional peripherals such as UARTS, timers, and other devices to be used to expand the capability of the microcontroller.
2. It provides a low-cost way of implementing systems that would otherwise need to use conventional microprocessors. The microcontroller can provide the bulk of the system's functionality on a single chip.
3. Expanded mode operation often makes the software development effort simpler and more efficient.

Generally, eight bit units lack some of the more sophisticated bus controls that are found on microprocessors. Microcontrollers normally do not have inputs to handle wait states, for example. They also must generally serve as the bus master and usually cannot give up the bus to other devices such as DMA controllers.

One of the first widely successful eight bit microcontrollers was Intel's 8048. This chip is still available, though it is not commonly used in modern designs. Architecturally, the 8048 represents a classic microcontroller as shown in Figure 10-1. The accumulator is 8 bits with 1K of program store and 64 bytes of RAM. The RAM is of the "extended register" set variety and cannot be used to execute code.

The instruction set is fascinatingly idiosyncratic, and only a true believer in the 8048 can really appreciate it. There is no subtraction instruction. Subtraction is accomplished by the more roundabout process of complementing the subtrahend and then adding. Index-style instructions are possible but only over a limited range of the address space.

The 8048 can be run in either the *expanded* mode or in a *stand-alone* mode. It was one of the first microcontrollers to offer this flexibility. The various modes of operation will be discussed shortly.

The success of the 8048 led to Intel introducing the extremely popular 8051 series of microcontrollers. The 8051, like the 8048, is an eight bit unit. It has an improved instruction set, serial I/O, larger program store, and RAM. One of the hallmarks of the 8051 is its ability to do Boolean operations on a bit-wise basis. This is a powerful capability, though it is somewhat awkward for the uninitiated to actually put into practice. Due to its early introduction, its availability in military temperature ranges, and its processing power, the 8051 has become a mainstay of military programs. It is indeed rare to see any microcontroller other than the 8051 used in military applications.

Motorola took a very different approach with its successful 6801 series, than Intel did with the 8048 series. Motorola adopted the 6800 core to form the heart of the 6801 device. As with its conventional microprocessors, Motorola's microcontrollers make use of fewer but more powerful accumulators. The RAM is of the true working memory type and can be used to execute code or to store data.

The power of the 6801 was demonstrated by the introduction of the L'ilbug version of the chip. This was the first microcontroller that could be used as a single chip, stand-alone, interactive computer. The on-board ROM of the 6801 was programmed with a monitor (see Chapter 12) that allowed code to be downloaded to the RAM via the on-board serial port. Once loaded, the code could be executed, complete with single step and break point options. The available on-board memory is so limited that the stand-alone mode is really just for show. In the expanded mode the monitor is a handy development aid, however.

Motorola has two other families of eight bit units. On the low end is the 6805 series. These devices are basically eight bit microcontrollers with a four bit mentality. The 6805 series is very popular in cost-sensitive applications requiring the use of eight bit data types.

On the high end is the 68HC11 series of devices. The 68HC11 is probably the most powerful eight bit microcontroller available today. Its sophisticated instruction set, combined with powerful onboard peripherals, supports advanced programs.

The 6801 and 68HC11 families are very popular in the automotive industry. The low cost, high performance, and ease of use of these units insure that Motorola will be well represented in Detroit for a long time to come.

The capabilities of the eight bit microcontrollers are improving continuously. Undoubtedly this trend will continue into the indefinite future. As we noted at the beginning of this section, eight bit microcontrollers are the most popular of the families. Unlike microprocessors, the pressure to achieve greater performance by increasing the bus width is not strong. Instead, performance is being increased by improved instruction sets, more sophisticated on-board peripherals, and larger memories.

It is likely that eight bit microcontrollers will continue to represent the optimum price/performance solution for the indefinite future.

10.6 Sixteen Bit Units

While the pressure to expand to 16 bit microcontrollers is not as strong as the pressure to expand to 16 or 32 bit microprocessors, there are still many applications that can benefit from the use of 16 bit systems.

Some applications always require more processing power than is currently available. This is particularly true in the automotive field where increased performance and functionality is a constant goal.

Currently, the most common microcontrollers with a 16 bit architecture are Intel's 8096 family. This family consists of machines with high-performance architectures, sophisticated analog inputs, and high-speed digital outputs. The high-speed digital I/O consists of specialized timing circuitry that is designed to optimize the pulse control sequences often found in automotive applications.

10.7 I/O Interfacing

We have already looked at how the synchronous serial I/O port can be used to expand the I/O of a microcontroller. Similarly, the on-chip asynchronous serial ports are quite straightforward. RS-232, RS-422, and other connections are quite easily implemented, and we will not elaborate on them. Instead, we will look at some of the more unique and interesting ways of connecting the microcontroller to the rest of the world.

First, we will look at expanding the I/O using conventional peripheral devices. As noted earlier, many microcontrollers have the ability to run in an expanded mode, in which they resemble conventional microprocessors. For example, Figure 10-6 shows two ways of adding an extra serial channel to a microcontroller. In Figure 10-6a we have connected the

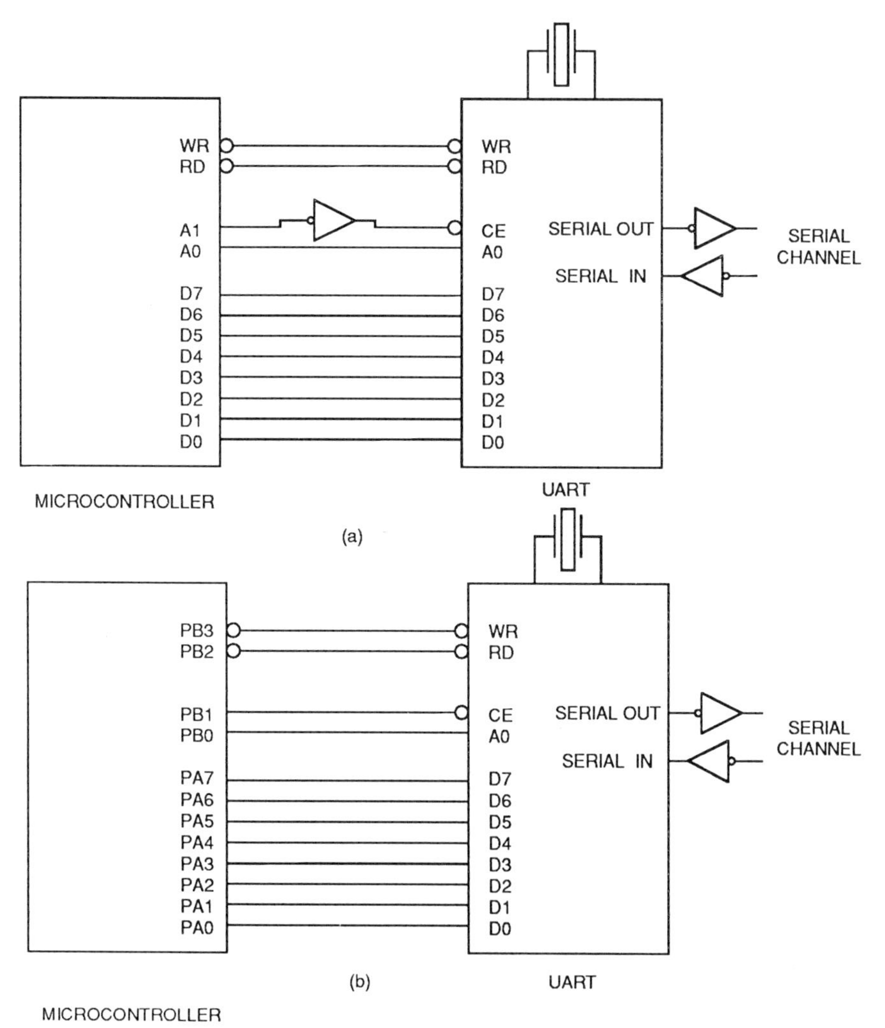

Figure 10-6
Microcontroller Expansion.

external UART using a multiplexed bus. The benefit of this approach is the program efficiency that can be realized. The UART is addressed as a standard part of the microcontroller's address space. Conventional instructions can be used to set up the UART and to transfer data. For example, the program:

```
UART    EQU     $10          ; Base address of the UART.
CTRREG  EQU      $1          ;  Offset for the control
                             ;   register.

START   LDAA #PRBITS         ; Get UART configuration bits.
        STAA UART+CTRREG     ;  Put them in the control
                             ;   register.

        LDAA #'A'            ; Character to be transmitted.
        STAA UART            ; Send it.
```

would configure the UART and transmit the letter *A*.

Figure 10-6b takes a different approach. In this case, the microcontroller is left in its nonexpanded mode (or one is being used that does not have an expanded mode capability.) The data bus of the UART is connected to the static I/O port A of the microcontroller. The chip enable of the UART (CE) is connected to the static I/O port B, bit 1 (PB1). The read line and write line are connected to PB2 and PB3, respectively. The address input of the UART (A0) is connected to PB0.

We can still communicate with the UART but it is much more cumbersome. For example, to write the configuration word into the UART requires the following steps:

1. Configure port A as an output (one instruction).
2. Load the configuration bits into an accumulator and output them to port A (two instructions).
3. Set PB0 to a 1; PB1 to a 0. This sets up the UART to receive a word (one instruction).
4. Set PB3 to a 0. This enables the UART (one instruction).
5. Set PB3 to a 1. This completes the transfer (one instruction).

This is more complicated than the entire previous program, so the obvious question is "why would anyone want to take such an approach?" There are several reasons. First, of course, a microcontroller may be specified that has no expansion mode. If this is the case, Figure 10-6b is the only option for expanding the microcontroller's resources.

If the microcontroller does have an expanded mode, we may still wish to use this more roundabout approach. The reason is that this technique makes use of fewer pins than does an expanded mode implementation. When a microcontroller is used in the expanded mode, the majority of the I/O pins typically are dedicated to assuming standard address and data functions. By using the method shown in Figure 10-6b, we can get by with dedicating only the minimum number of pins to the I/O expansion. This is particularly true when interfacing to four bit peripherals such as real time clocks (even more so since these devices are not accessed very often).

Using conventional peripherals is only one way to interface to a microcontroller. It is often said that even if computers are *digital,* the world is *analog*. As microcontrollers find their way into more and more consumer products, the need to sense analog values increases. One simple approach to measuring analog values is shown in Figure 10-7.

The operation of the circuit is quite straightforward. The analog value to be measured is input to the voltage-controlled oscillator. The microcontroller simply measures the period of the signal coming out of the VCO. This period will be directly proportional to the analog input.

Measuring analog voltages is becoming such a common requirement that even many of the lower priced microcontrollers are being equipped with built-in ADCs. Typically, these converters have built-in sample-and-hold circuitry and are very easy to use.

Typical conversion times for built-in ADCs are in the 30 to 100 μsec range. That is not fast by flash ADC standards, but these times are generally fast enough for most practical applications. The resolution of these converters is generally eight bits. A common feature is a multiplexer that allows more than one channel to be sampled. Four to eight input multiplexers are typical.

Often times a microcontroller application does not require knowledge of the exact magnitude of a voltage. It is only of interest whether or not the monitored voltage is above or below a certain value. Monitoring the status of a battery is a typical application. The battery voltage will vary slowly, and we are not concerned with what the actual voltage is at

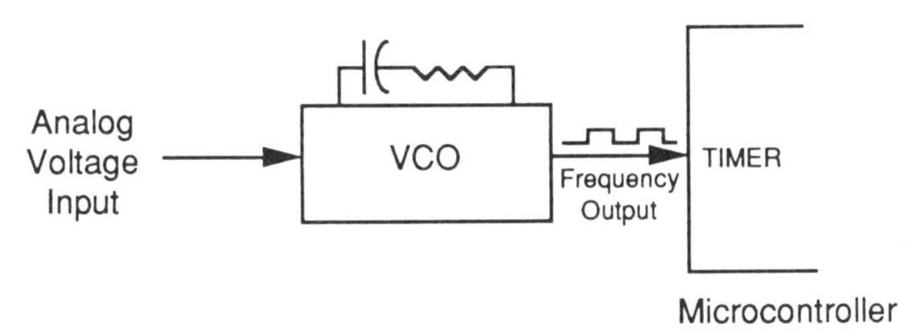

Figure 10-7
VCO-Based Analog-to-Digital Conversion.

any given moment. The signal is only of interest when the value falls below a safe operating level.

Peak detection is another such situation. In process control applications we often want to respond when a given parameter exceeds a certain value. But once this threshold has been exceeded and we have acted, we may want to increase the threshold to avoid a constant "tripping" of the alarm.

A low-cost solution to these requirements is the variable threshold input. A typical implementation of a variable threshold input is shown in Figure 10-8. The basic idea is simple. The voltage to be monitored is input to one side of a comparator. The other side of the comparator is connected to a DAC. The desired threshold is set by adjusting the DAC. When the input voltage is above the threshold voltage, the comparator's output will be read as a one; when the voltage is below the threshold, the output will be read as a zero.

Given enough time, the microcontroller can use this technique to get a good approximation of the magnitude of the input voltage. By adjusting the DAC until a change in the least significant bit causes the comparator to change states, the microcontroller can "hunt" for the magnitude of the voltage. In fact, this is usually how most complete ADCs work. This technique is called conversion by *successive approximation*.

Low-powered DACs such as these are fairly easy to put on a microcontroller. Scaling up the output power to a level that is usable as an output signal is a formidable problem, however. The area required for the analog drive is large. Further, shielding the sensitive digital circuitry from the noise and heat of the analog section is not easy.

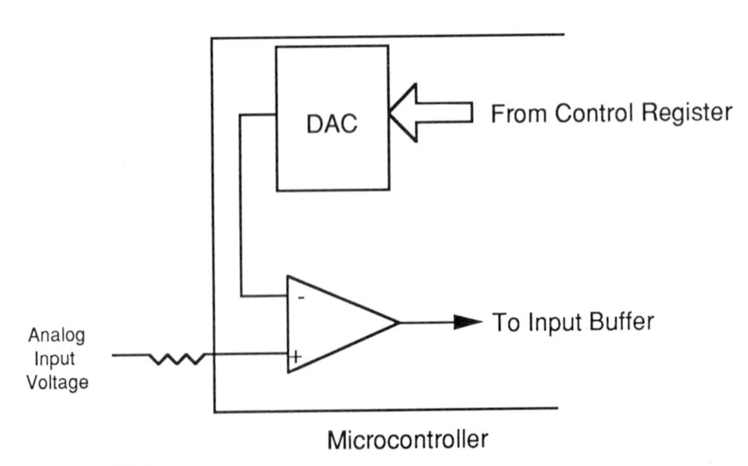

Figure 10-8
Programmable Threshold Input.

Fortunately, there are often alternatives to true analog outputs. In some cases, these alternatives are actually preferable to true analog outputs. A case in point is *pulse width modulation* (PWM). As an example, let us look at a low-cost, motor speed control application. Figure 10-9a shows an analog approach to controlling the speed of a motor. The motor is the inexpensive DC brush-type commonly found in toys, portable hand-tools, etc.

Figure 10-9b shows an equivalent circuit for the analog control. To provide effective control of the motor's speed, R_Z must range from 0 Ω up

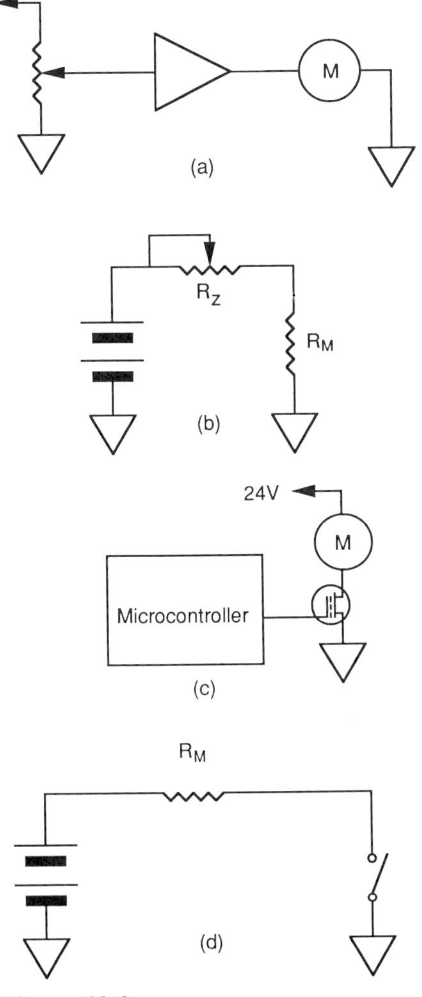

Figure 10-9
Analog versus Digital Power Control.

to approximately $5R_m$. Let us consider the efficiency of such a circuit. At the low end, when $R_z = 0\ \Omega$, the efficiency is good since all of the power will be dissipated in the load R_m. At the other extreme, when $R_z = 5R_m$, the overall current through the circuit will be relatively small and very little power will be dissipated, so again efficiency is not bad.

Now consider the case when $R_z = R_m$, which is the the midway point in the control range. In this configuration, R_z and R_m form a classic divide-by-two voltage divider. Half of the power will be dissipated in R_z, and half the power will be dissipated in R_m. Or in other words, half of our available power will be wasted in the control circuit rather than doing actual work.

Figure 10-9c shows a digital approach to the problem. The microcontroller is attached to a transistor switch that is used to turn the motor on and off. The equivalent circuit is shown in Figure 10-9d. When the transistor is off, it is as if the switch were open. No current flows, so no power is lost. When the switch is closed, all of the voltage drop is across the load R_m. Therefore, all of the power is dissipated in the motor. No energy is lost to the control circuit.

But what about speed control? This is achieved by turning the motor on and off very rapidly. Obviously, for full power to the load, the transistor is left turned on continuously. For no power to the load, it is left turned off. By setting some ratio of on-time to off-time an effective power control anywhere between the two extremes can be achieved. By varying the duty cycle, the motor will respond *as if* the input voltage were being varied, but without the attendant inefficiencies of an analog control circuit.

Actually there are other benefits besides the power efficiencies. Since the motor sees a full voltage across its terminals whenever power is on, torque is improved. Most everyone who has played with an electric train set has experienced the phenomenon of clogging or stalling. This occurs when power is slowly supplied to a permanent magnet-type motor. What happens is that the motor moves just enough to effectively short out the strators. Once this situation occurs, applying more power only makes the situation worse. Eventually this can burn out the motor.

The high torque provided by a PWM control scheme helps to avoid this phenomena. The microcontroller can do even more, however. For example, on start up the motor can be given full power for a short period of time. Once the motor has overcome the initial friction and has started to build up momentum, the power can be eased back by beginning the modulation. This technique allows the motors to start up smoothly even when the target speed is very slow.

Pulse width modulation techniques can be applied to any load that has a high effective inertia. The technique works well for filament-based light bulbs, for example. Interestingly enough, it also works for fast-responding light sources such as LEDs. In the case of the LED, the diodes will typically respond as fast as the modulation. However, the eye will average the light coming from the device. The net effect is that the LED will appear to grow lighter or darker as the duty cycle is varied.

This is a particularly handy technique to remember when designing hand-held battery operated instruments. Often, effective power management can be obtained at only the cost of a discrete transistor and a resistor or two.

10.8 Chapter Summary

In this chapter the basic architectures of the commonly available microcontrollers have been presented. We have also examined some of the ways these devices can be connected to the real world. A few lines of assembly code were used to illustrate some points, but in general we have stayed away from the subject of software. In the next chapter we will examine the software architecture of these interesting and useful devices.

- Microcontrollers bring all the flexibility of a general purpose computer to a single device.
- Microcontrollers are typically found in one of three sizes: 4 bit, 8 bit, or 16 bit. Occasionally, specialized units are found in other sizes, typically 1 bit or 32 bit.
- Eight bit microcontrollers are the most popular. These units generally represent the best cost/performance option for general purpose applications.
- For consumer applications four-bit units are common. These units are used in cost-sensitive applications where the higher development costs are amortized over large volumes of products.
- A wide range of options are available for the program stores on most microcontrollers. EPROM or EEPROM for development, and one time programmable EPROM or masked ROM for production environments.
- EEPROM is becoming a common option on microcontrollers. EEPROM allows adaptive systems that learn over time and is also useful for storing system diagnostic information.
- Many microcontrollers can be used in an expanded mode. In the expanded mode, these devices function in a fashion similar to conventional microprocessors. This makes interfacing to standard peripherals easier and can simplify the product development cycle.

11

Micrcontrollers and Software

Now that microcontrollers have been examined from a hardware perspective, it is time to look at how the hardware is told what it is that it should be doing. Software generation for a microcontroller is fundamentally different from our previous software generation efforts. This chapter will look at some of the differences and also provide some insight into the more sophisticated programming techniques supported by several of the higher performance microcontrollers.

11.1 Patterns, Microcode, and Object Code

In the case of combinatorial PLDs, some might argue against calling our programming efforts software development. It could be claimed that we were only configuring the device not actually *programming* it. This is making a stand on a semantic difference, something that the author has strived hard to avoid in this book. But there are some interesting points that should be brought up.

In the source files for PLDs, there was no implication about when something happened based on its location in the file. That is to say if an equation were at the bottom of the file, there was no implication that it was executed last. In the case of combinatorial PLDs where everything happens simultaneously, this only makes sense. But it is also true for the microcode and state machine examples. The next state was always explicitly specified. This is not the case when programming a microcontroller.

executes the first instruction, then the second instruction, etc. This may seem obvious but the advent of sequentially executed object code was a fundamental advancement in computer science, the likes of which has not been seen since.

While high-level languages are growing in popularity, most programming for microcontrollers is done in assembly language. The purpose in this chapter is not to teach assembly language programming. Most of what one learns in studying assembly language for a microprocessor can be translated to microcontrollers. There are a number of books available on assembly language programming for microprocessors so the effort will not be duplicated in this text. Instead, the goal is to present otherwise hard-to-obtain information on some of the more sophisticated and misunderstood aspects of programming a microcontroller, specifically the areas of interrupts and real time multi-tasking.

In order to get to these goals in a meaningful way requires the laying of some groundwork. This groundwork will take us through the fundamentals of assembly programming at a rather fast pace and not in great depth.

11.2 Elementary Instructions

This section discusses a small subset of the instructions typically available in a microcontroller. The emphasis is on how these instructions are used to manipulate the hardware resources of the machine.

The upcoming examples will use the register model shown in Figure 10-2. The first register of interest is the program counter (PC). The PC is the mechanism that actually controls the program flow.

The PC points at the instruction to be executed. As the instruction is being executed, the PC is incremented to point at the next instruction. As one can imagine, the PC plays a pivotal role in the software process. When the program is progressing normally, the programmer is not concerned with the PC. It automatically increments to the next instruction. However, it is sometimes necessary to execute some instruction other than the next one in the sequence. This is accomplished by loading a new va and the address of the instruction into the PC.

There are two basic types of load instructions: *direct* and *indirect*. T can have different names for different architectures, such as immec nd indexed. As the name implies, a direct instruction gets its value directly from the code. For example, a typical instruction might look like this:

```
ADDRESS    CODE     MNEMONIC          COMMENTS
0000      01 01 00 JMP #$100 ; Jump to address 100.
```

The *address* is simply where the instruction is in memory. The *code field* contains the bytes that actually make up the instruction. These are generated by the assembler from the *mnemonic field.*

The first numeric value of 01 in the code field is called the *opcode* (operations code) for the mnemonic JMP. Whenever the assembler sees "JMP #" it substitutes 01H. The next two bytes are the address of the instruction to which control is to be transferred.

The mnemonic is called the *source code* for the instruction. The # sign preceding the value of 100 tells the assembler that the addressing is immediate (i.e., direct). The $ indicates that the value is expressed in HEX.

The other information is there only to make reading the program easier for a person. Since only the bytes in the code field are actually stored in the microcontroller's program store, truly hard-core programmers can code the machine program directly, and they do not even need the assembler to translate the mnemonic field. Such programs are impossible to maintain or modify however, and the practice is discouraged even for those who can do it. (Doing direct programming like this can be a valuable training exercise and is somewhat fun for the hard-core among us.)

Now, back to the instruction itself. What the JMP instruction actually does is to load the address given in the instruction directly into the PC. In this case, the next instruction the program executes would be at location 0100H.

The second type of addressing, indirect, is similar. However, in indirect addressing the address field of the instruction is the *address* of the value to load, not the actual *value* to load. This is shown in the following code:

```
0000   02 02 00 JMP $200   ; Indirect Jump @ 0200.
                ORG $0200  ; Set the address to 0200.
0200      FO 00 FDB $F000  ; Address to execute.
```

Notice that the # sign has been left out of the instruction. This is how the assembler knows to generate a 02 opcode instead of a 01 opcode. What happens in this case is that the program counter is loaded with the value *stored at* address 0200H and will then *execute* the instruction at address F000H.

Indirect jumps are handy in a variety of programming situations. A table of commands are often built up. The address of the instructions to be executed for each command is stored in the table. The command is executed by simply looking up the address in the table and transferring control to the stored address. Programs based on such command tables are called *table driven*.

An important use of indirect addressing is in the hardware/software interface. Consider what happens when a microcontroller gets a hardware reset, for example. Program execution must start *somewhere*. For devices in the 6800 family, this address is stored at address FFFEH and FFFFH, the last two bytes in the memory map. The program can be anywhere in the 64K memory space, but the address of the first instruction must be stored at location FFFEH and FFFFH.

The reset function simply transfers the contents of the last two bytes into the PC. The address transferred is called the *reset vector*. As will be shown shortly, other hardware functions also make use of other fixed locations for storing vectors.

Other instructions exist for loading the registers (LDAA for load accumulator, LDX for load index register, etc.). As one might expect, the reverse is also true. Store accumulator A (STAA) puts the contents of the A accumulator into the address specified in the instruction. These operations can be combined with logical function as well. For example:

```
0000      02 0F ANDA #$0F ; Clear High nibble.
```

This will set the upper four bits of the accumulator to 0000_B. The lower four bits will remain exactly as they were. These instructions will be illustrated shortly.

Before talking about the next instruction of interest, the operation of the stack pointer (SP) needs to be covered. As the name implies, this register *points* to the *bottom of the stack*.

The operation of the stack is quite simple. Pushing something on the stack causes the following to happen:

1. The value to be pushed is stored in memory at the location to which the SP points.
2. The stack is then decremented by the size of the value stored on the stack.

For example, if the stack contains the value 0100H, and the index register (X) contains the value FFAAH, then pushing X on the stack will cause the following:

- FFH is stored at 0100H.
- The SP is decremented to 00FFH.
- AAH is stored at 0FFH.
- The SP is decremented to 00FEH.

This sequence is illustrated in Figure 11-2. Figure 11-2a shows the SP and the memory before the push. Figure 11-2b shows the SP and the memory after the push. Popping something from the stack simply reverses the sequence.

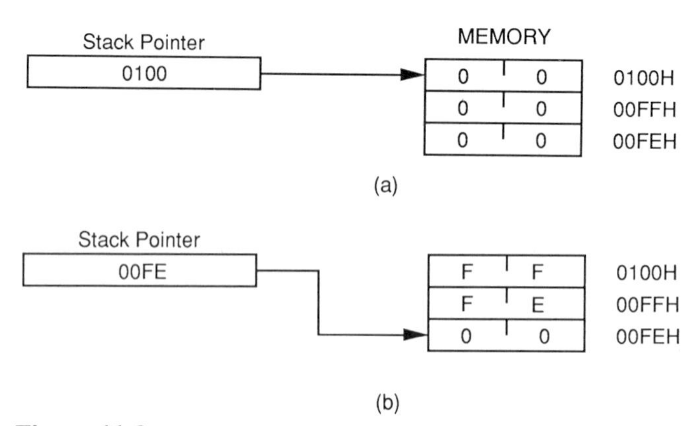

Figure 11-2
The Stack in Operation.

One of the most common uses of the stack is with the jump to subroutine (JSR) instruction. The JSR instruction is similar to the JMP instruction except that the current contents of the PC are first pushed on the stack, and the PC is loaded with the address of the instruction to execute. The last instruction executed by the subroutine is a return from subroutine (RTS). This instruction simply pops the address from the stack back into the PC. The program will continue executing the instructions immediately following the JSR instruction.

The final class of instructions to discuss are the *conditional jumps*. These instructions are often called *branches*. Branches generally make use of *relative addressing*. Relative addressing adds or subtracts from the PC rather than simply loading a new value.

Branches form the decision-making capability of the microcontroller. A good example is the branch on zero (BRZ) instruction. A typical program sequence might look like this:

```
0100 03 01       ANDA #$01 ; Test bit 0.
0102 04 04       BRZ  SKIP ;
0104 05 00       LDAA #$FF ; Set A to all zeroes.
0106 07 02       BRA  NEXT ; Continue.
0108 05 20 SKIP  LDAA #$00 ; Set A to all ones.
010A       NEXT  ....
```

The purpose of this section of code is to set the register to all zeroes if bit 0 is a 0 and to set it to all ones if bit 1 is a 1. Here is how it works:

- The ANDA operation will set the Z bit of the CC register if the result of the AND is a zero. Otherwise, the Z bit will be cleared.
- The BRZ instruction checks the Z bit.

Case 0. If the Z bit is set, then 4 is added to the PC. This will cause the PC to skip the next two instructions. The LDAA #$00 instruction will be executed, loading all zeroes into the accumulator. The program will then proceed.

Case 1. If the Z bit is cleared, then the BRZ instruction will do nothing. The LDA #$FF instruction will be executed next, setting the accumulator to all ones. The branch always (BRA) instruction will add 2 to the PC. This will skip the LDAA #$00 instruction. The program will proceed.

Most microcontrollers have a wide variety of conditional jumps to test for various states. As shown in this example, the term *conditional jump* can be loosely applied to an instruction like BRA. BRA has the same form as the true conditional jumps, but since the branch is always taken, it is not truly conditional. An even more interesting "conditional jump" is the branch never (BRN) instruction. As the name implies, the branch is only taken if 1 AND 0 = 1, which of course is never.

It may seem silly to have a conditional instruction that can never find a true case for the branch, but it actually does have some uses. One use is as a debugging tool. A conditional branch instruction can be temporarily replaced with a BRN instruction to effectively deactivate the branch.

These are only a few of the instructions found in a typical microcontroller. The exact flavor and style will vary from family to family and sometimes even within certain families. The next section will look at how these instructions are put to use.

11.3 Simple Program Structure

Figure 11-3 is a conceptual model of a microcontroller from a software point of view. On the left-hand side is an input port (port A). Normally, ports are eight bits wide, and we will assume that all of our ports are such. The input port is connected to the accumulator, the RAM, and output port B. The path that connects these units is the data bus.

All traffic is conducted through the accumulator. In our model, the accumulator will participate in any data operation. This is the case in the real world for most microcontrollers. Microcontrollers with a register-to-register architecture are occasionally encountered, but these are the exception. And even with these machines, the basic sequence of data moves to be discussed remains the same.

To get a feel for how a microcontroller works, let us look at a simple problem. Notice in Figure 11-3 that one of the inputs (bit 0 of port A) is

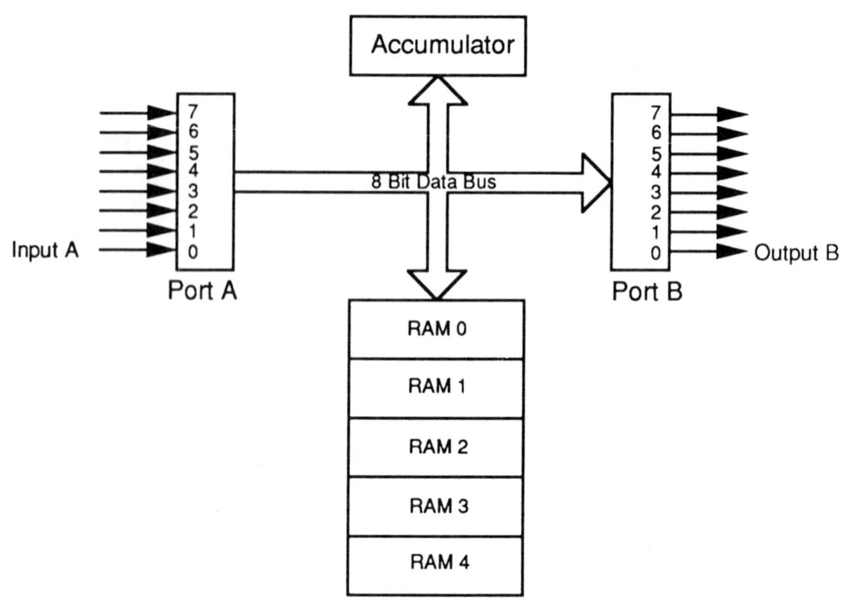

Figure 11-3
Conceptual Programming Model.

labeled A, and one of the outputs (bit 0 of port B) is labeled B. The simplest logical operation is to set B = A. One program to accomplish this is:

```
LDAA PORTA ; Get the data on Port A.
STAA PORTB ; Send it to Port B.
```

If this program is executed, the bits on port A would be read into the accumulator. The accumulator would then write these bits to port B. Thus, B would be set equal to whatever was on A.

This simple program does what we wanted but it has a few limitations. First, it only sets B equal to A *once*. For the second limitation, this simple program sets *all* of the bits on port B equal to *all* of the bits on port A. This may be acceptable (it meets our original specification), but we normally would only want to update B and leave the other seven bits alone.

The first problem can be solved by making the program into a loop:

```
START    LDAA PORTA; Get the data on Port A.
         STAA PORTB; Send it to Port B.
         BRA  START; Keep doing it.
```

In this case, the program simply passes control to itself *ad infinitum*. This loop will continue as long as the microcontroller has power, so whenever *A* changes, *B* will be updated. This infinite loop, which most computer programmers are taught to avoid at all costs, is a very common occurrence in microcontrollers.

This program still has the problem that *all* the output bits are changed, however. This can be corrected by using the logical operations that are possible with the accumulator. For example:

```
TEMP    RMB         ; Define 1 Byte of storage.

START   LDAA PORTA; Get the input bits.
        ANDA #$01  ; Clear all bits except bit 0.
        STAA TEMP ; Save the image in RAM.

        LDAA PORTB; Get the output bits.
        ANDA #$FE ; Clear bit 0.

        ORAA TEMP ; Combine the two images.
        STAA PORTB; Output the image to Port B.
        BRA  START; Form a continuous loop.
```

In this example one byte of RAM, labeled TEMP, has been reserved to act as a temporary storage register. Next, the accumulator is loaded with the data on port A, as before. This time, however, the contents of the accumulator are ANDed with a constant hexadecimal value of 01. The effect of this is to set all bits, except bit 0, to 0. Bit 0 will be *conditionally* set: a 0 if A was a 0; a 1 if A was a 1. Then this image 0000 0001B or 0000 0000B, depending on what the input was, is saved in RAM.

The next part of the operation is to get the *output* bits loaded into the accumulator. We again AND the bits, this time with the bit pattern 1111 1110. Bits 7 through 1 will be unaffected by this; bit 0 will be unconditionally set to 0. In other words, the accumulator now has an image of what is on port B, except that bit 0 is clear.

The ORAA TEMP instruction ORs the two images together. This new image is identical to the current bit settings of port B, except that bit 0 is now set to whatever was on the input bit 0 of port A. Finally, this image is output to port B. This program will cause B and only B to equal A.

This is a rather artificial situation of course. But this exact situation does show up from time to time. More important, however, is the way this program demonstrates the use of logical instructions to accomplish the bit-wise manipulation of an output port. This is often the preferred technique not only in external I/O but also when updating internal registers such as the interrupt control register.

Some other interesting insights can be gleamed from this program as well. This program will use about 15 bytes of memory, depending upon the microcontroller. The speed of execution is also highly dependent upon the microcontroller, the oscillator frequency, etc. Normally, a program like this would need about 30 μsec to execute one complete loop. This is extremely fast by human standards but quite slow in the electronics world.

By contrast, a PLD programmed for A = B would insert around 20 nsec of delay between the input and the output. This order of magnitude difference in speed clearly shows why PLDs are used in high-speed memory address decoders and microcontrollers are not even considered for such applications.

Let us look at another application. Figure 11-4 shows a simple block diagram of a microcontroller used to implement the basic push button counter first introduced in Chapter 2. The microcontroller reads the status of the switch and maintains a count of the number of times the switch goes through an open–closed–open cycle. Figure 11-5 shows a flowchart for one way in which this function could be realized.

The operation of the counter is straightforward. At power on, a simple power-on clear (POC) circuit resets the microcontroller. In this case, the POC is a simple RC combination, generally chosen to give a time constant in the area of 10 msec.

When the microcontroller comes out of reset, it will do an indirect jump via the reset vector to the start of the program. The first thing the program does is the initialization. Port A, bit 0 is programmed as an input, and port B's bits are programmed as outputs. Then the program clears the variable called "count."

The program will read the input port next. At this point, it will check the value of the bit that is read in on PA0. If the switch has not been pressed, the bit is a one and the program simply loops. If the button is then pressed, the input bit will be a zero and the program continues to the next step. This process of reading the switches is called *polling*.

At this point the program delays for 20 msec. The purpose of this delay is to allow the switch time to settle (see the discussion in Chapter 2 on switch debouncing).

The input port PA is read again. This time it is tested to see if the input bit is a zero. If it is, the switch is still being pushed down and the program loops. When the switch is released, the input bit will be a one and the program will continue.

This second test is necessary to prevent a "run away" condition in which the count is incremented *while* the switch is closed, rather than being incremented *once each time* the switch is closed.

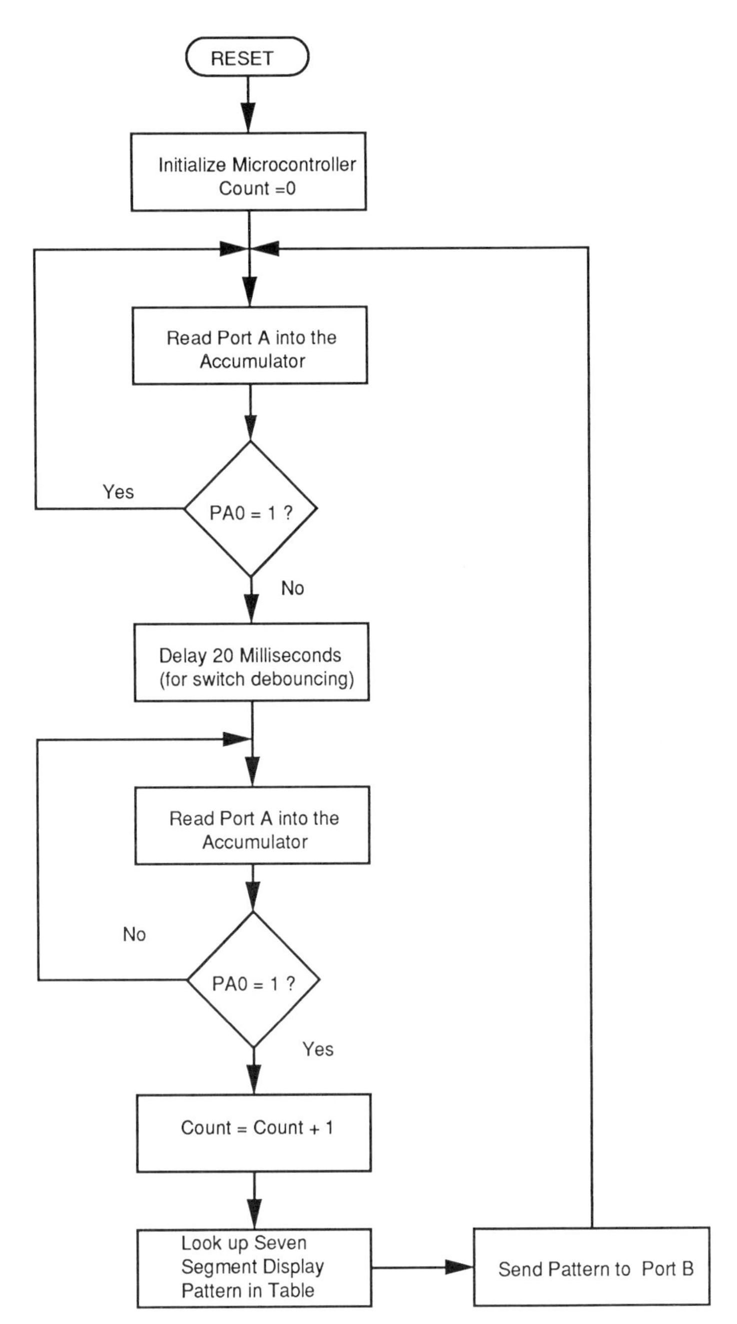

Figure 11-4
Push Button Counter Algorithm.

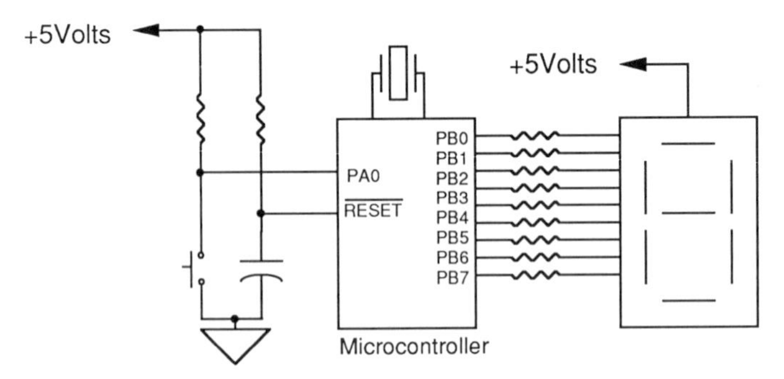

Figure 11-5
Microcontroller-Based Counter.

A positive closure of the switch has been confirmed so the count is incremented by one. Some microcontrollers have a special instruction that will increment the variable count in one instruction:

```
INC COUNT;          Increment the count variable.
```

Others require adding a 1 to the variable:

```
LDAA COUNT  ; Get the variable.
ADAA #01    ; Add 1 to it.
STAA COUNT  ; Save the variable.
```

After adjusting the count, the output pattern for the seven segment LED display is looked up in a table. This is commonly done by using an indexed form of the load instruction. For example, a typical sequence might look like this:

```
LDX  TABLE ; Get the address of the lookup table.
ADDX COUNT ; Add in the offset for the count.
LDAA X     ; Use the X register as the address.
```

The display pattern is then sent to the output port B where the proper segments are lighted for the current count.

This is a simple program but it demonstrates some very important concepts. It is easily seen that the functionality of the counter depends almost entirely on the software. Functions such as the contact debouncing of the switch are handled in the software. This ability of the software to define the functionality provides great flexibility.

If a stopwatch were desired instead of a counter, for example, the hardware would not need to be altered. The same circuit could be used for a counter or a stopwatch, with selection of the use made by choosing which software is installed. Features can be added to the circuit without impacting the hardware design.

Examining the structure of the program also shows some interesting things about applying microcontrollers. If we look at the amount of time the microcontroller spends performing each step, an interesting pattern emerges. The start-up sequence requires only a few microseconds to clear the variable. Likewise, the actual incrementing of the count typically requires less than 10 μsec. Most microcontrollers have instructions for looking up entries in tables, so the conversion of the count to an output pattern for the seven-segment display will usually take only a few microseconds.

Most of the time is spent in one of three places: The first switch-closure test loop, the 20 msec delay, or the second switch-release test loop. Notice that none of these are directly related to the actual process of counting. These operations are simply *overhead* functions used in our particular implementation of the counter. If all we want (or need) the microcontroller to do is act as a counter, these overhead activities represent no loss. On the other hand, if there are other useful things the microcontroller could be doing, this is a waste of resources.

Consider the case with the switch test loops: Even if a human operator is pressing a mechanical switch rapidly, the microcontroller will look at the switch hundreds of times between each switch closing. It is not hard to imagine scenarios where the microcontroller might spend 99% or more of its time in overhead functions and only 1% of its time doing actual work.

This low utilization of the processors capabilities has led to the development of a variety of software and hardware optimization techniques. Some of these techniques can be illustrated with a microcontroller-based print engine in a dot matrix printer. A simplified diagram of such an application is shown in Figure 11-6.

This printer is configured to accept parallel data in the popular Centronix format. The ASCII code for the character to be printed is placed on port A. The device driving the printer (typically a computer) will then drive the STROBE line low. The printer acknowledges that it has received the character by pulsing the ACK (acknowledge) line low. To show how a microcontroller might be programmed in such an application, here is part of the code:

```
LOOP LDAA PORTB ; Get the input bits.
     ANDA #$01  ; Isolate bit 0.
     BNZ  LOOP  ; Wait for it to go low.

     LDAA PORTA ; Get the character to print.
```

```
(Move the motors, fire the print heads, etc.).

LDAA PORTB ; Get the control bits.
ANDA #$FD  ; Set bit 1 low.
STAA PORTB ; Set ACK low.
ORAA #$02  ; Set bit 1 high.
STAA PORTB ; Set ACK high.
BRA  LOOP  ;
```

This program manipulates the bits in the same manner as the examples earlier in the chapter. Notice, however, that a conditional loop is formed by the use the BNZ (branch not zero) instruction.

The AND instruction preceding the BNZ instruction is used to set the Z bit. If STROBE is high (that is, a character is not ready on the input), the result of the AND operation will be 0000 0001B. The Z bit will be cleared and the loop will be taken. Thus, the program will "wait" until it reads a zero on the strobe line. When it does read a zero on the input, the program loads the character on port A into the accumulator. It then goes through the sequence of stepping the motors to position the print head, fire the print heads, etc. until the character has been printed.

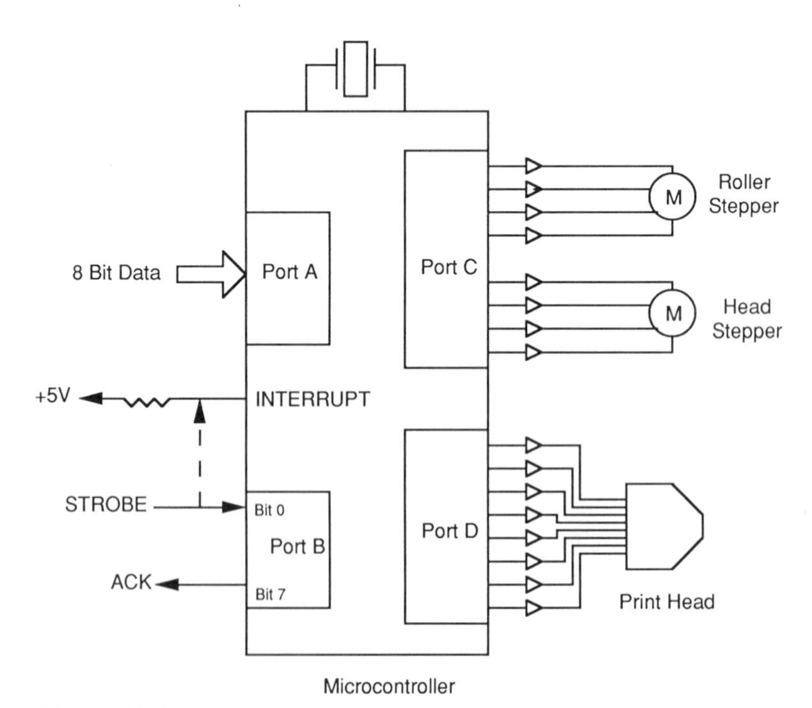

Figure 11-6
Microcontroller-Based Dot Matrix Printer.

Once the character has been printed, the ACK line is pulsed low to tell the driving device that the printer is ready for another character. This sequence continues over and over while the printer is in operation. It is this "wait until ready, then do . . ." kind of operation that make micro-controllers extremely useful.

In the last section the jump to subroutine (JSR) instruction was introduced. The next section will demonstrate how this instruction can be used to improve the coding.

11.4 Subroutines and Program Structure

The previous dot matrix controller program was fairly straightforward. Things tended to follow a smooth sequence of steps through the loop. Still, it can be seen even from this simple example that it could get difficult to follow when one major block of the code was beginning and the previous block ending.

For a real world program this problem can become significant. Long stretches of code make it difficult to tell what is actually happening. This leads to confusion, and changing one instruction can have unexpected consequences further down the line.

The solution to this problem is to add *structure* to the program. While much time has been spent (read: wasted) in computer science debating exactly what constitutes structure, the working definition is simple: structure is a matter of partitioning the program into manageable logical blocks.

In a structured program each block should have well-defined inputs and outputs. No block should make use of some obscure event in another block; all information should be explicitly passed as inputs or outputs. As a general rule, each block should be no longer than one page and at the absolute most two or three pages. If a block is longer than this, it is likely that too much is going on. The code should be partitioned into smaller, more manageable pieces.

Let us look at a structured version of the dot matrix control program as shown in Figure 11-7.

The major functions have been partitioned into logical blocks. The actual code for GETCHR, for example, is identical to the previous example. However, now it has been broken out into a subroutine.

Notice how this improves the readability of the program. At the main program level it is obvious what we are trying to do. There are no details of implementation getting in the way of understanding what the program is to do.

```
************************* Main Program. *************************

       LOOP      JSR  GETCHR      ; Get the input character.
                 JSR  PRINTIT     ; Put the character on the paper.
                 JSR  CHRACK      ; Acknowledge receiving the character.
                 BRA  LOOP        ; Keep doing it.

************************ Supporting Routines. ******************

       GETCHR    LDAA PORTB       ; Get the input bits.
                 ANDA #$01        ; Isolate bit 0.
                 BNZ GETCHR       ; Wait for it to go low.

                 LDAA PORTA       ; Get the character to print.
                 RTS              ; Return to main routine.

                 PRINTIT  (Move the motors, fire the heads, etc)
                 RTS              ; Return to main routine.

       CHRACK    LDAA PORTB       ; Get the control bits.
                 ANDA #$FD        ; Set bit 1 low.
                 STAA PORTB       ; Set ACK low.
                 ORAA #$02        ; Set bit 1 high.
                 STAA PORTB       ; Set ACK high.
                 RTS                                   ; Return to main routine.
```

Figure 11-7
Structured Assembly Program.

Each subroutine is also an independent block. It is clear what the subroutine is trying to do, and one can be somewhat confident that any changes will be localized to the individual routine. Each routine ends with a return from subroutine (RTS) instruction. This simply transfers control back to the main program immediately after the JSR instruction that called the routine.

The use of subroutines conveniently modularizes the program. Subroutines offer other advantages as well. The code in a subroutine is reusable. For example, our printer may have several modes of operation (normal printing, landscape printing, graphics, etc.). Each mode will need some common resources such as getting characters from the input port. By coding the input routine as a subroutine, each of these modes can call it. This saves duplicating the input code for each routine.

Common operations, such as getting characters and making conversions, can be coded once for a variety of applications. These subroutines can then be incorporated into the program whenever needed, thus eliminating the need to keep "reinventing the wheel."

The actual implementation of subroutines may vary. For example, with the more sophisticated microcontrollers the stack can be set anywhere in the memory space and can be as deep as wanted. With the simpler microcontrollers the stack may be limited to storing as few as four addresses. With these units, the SP is a simple two bit counter.

Often the SP cannot be set and can only be cleared. With these types of units, the stack can only be used to support subroutines and interrupts. With the more general purpose units, the stack can be used for passing variables, temporary storage, and other such uses, as well as handling return addresses.

Subroutines are not entirely a blessing. In the above example, valuable structure was gained but at the cost of another valuable resource: time. Each JSR and RTS instruction adds execution time to the program. This increased execution time must be weighed against the benefits of reusable code and the improved maintainability of structured programs. When programming mainframes, minicomputers, or even PCs, the benefits of structured programming will virtually always outweigh the small impact on performance. Microcontroller programmers do not always have this luxury, however. Small code spaces, scarce RAM, and an absolute need for speed are common problems facing the microcontroller user. The need for maximum performance may force a compromise in the degree of modularity and structure that a program can support. It is this need for optimal performance that drives the discussion of the next structure: the interrupt.

11.5 Interrupts

Interrupts are similar in architecture to subroutines. The major difference is how they are called. Interrupts are triggered by hardware. In effect, an interrupt routine allows a hardware event to call a subroutine to handle its needs. First a discussion of how this is done is required. Then we can look at why an interrupt is a handy feature.

When an interrupt event is triggered, the microcontroller performs the following actions:

- The processor completes the instruction it is executing.
- It pushes the address of the next instruction onto the stack (just as in the case of a subroutine).
- Key registers (typically the condition code register and sometimes all of the registers) are pushed onto the stack.
- It then sets the interrupt mask bit in the condition code register. This prevents any other interrupt from interrupting the current routine.
- Then it does an indirect jump to the interrupt routine.

Most microcontrollers specify a certain address for an interrupt. Let us assume that the address for an external interrupt is $FF00_H$. If the interrupt routine is located at address $0C37_H$, then $0C37_H$ would be stored at address $FF00_H$. When the interrupt occurs, the value $0C37_H$ will be loaded into the program counter. The address of the interrupt routine, $0C37_H$ in this case, is called the *interrupt vector*.

The interrupt routine is terminated with a return from interrupt (RTI) instruction. The RTI is similar to the RTS instruction except that all affected registers, including the condition code register, are restored. The RTI simply pops all of the values stored on the stack back off the stack and into the respective registers. To the interrupted program the net effect of all of this is that the program execution continues just as if nothing had happened.

To see the value in this, let us consider the situation with the printer. When a character is input, the receipt is not acknowledged until we have actually finished printing the character. This insures that no character is sent before we are ready for it. This also means, however, that the driving device is tied up while we do the slow mechanical operations. It would be much better if the microcontroller took in the character, acknowledged it, and *then* did the printing. But this would only help a little.

The first character would be acknowledged immediately, and then actual printing would begin. The driving device could immediately send the next character, but the program could not do anything with it. The microcontroller would be moving the motors, firing the print heads, etc. The program could not take in the character until it got back to the top of the program loop. This can be a particularly bad problem if the system is involved with long operations such as generating a form feed.

There is another limitation to this approach. The constant polling of the STROBE line takes up valuable processor cycles. In this example, it does not really make any difference since there is not anything to do until the next character is received anyway. But if there *were* other things we could have been doing, this would be a problem.

Now, let us see how the interrupt line helps. If the STROBE line is connected to the external interrupt line instead of bit 0 on port B, the result would be an *interrupt driven* system. The interrupt vector is set to point to the GETCHR routine. When the driving device signals that a character is available, the microcontroller will stop what it is doing and execute the GETCHR routine.

This frees the program from having to wait for each character. Once the system gets the first character, it can acknowledge it and release the driving device. Simultaneously, the program can start printing the character. If a new character is sent before the first one is finished printing, the GETCHR routine can simply save it to a queue. When the system is done printing the first character, the next character is fetched from the queue for printing.

Notice that from the outside perspective it appears that the printer is actually doing two things at once. It is printing characters at the same time that it is accepting characters from the driving device. This is the beauty of interrupts; they free the microcontroller to rapidly respond to real time events.

Finally, one tangential point needs to be made. Interrupts are by definition hardware events. However, over the years it has become popular to add instructions to microcontrollers and microprocessors that simulate the interrupt triggering sequence. Originally, these were put in simply to make it easier to debug hardware interrupts. These software interrupts proved useful in a variety of debugging situations also. Their use has grown to the point where some operating systems, such as Microsoft's PC-DOS, use software interrupts for all of their function-calling conventions.

The difference between a hardware interrupt and a software interrupt can be confusing, and care must be taken to insure understanding of which is being discussed. In microcontrollers, software interrupts are generally limited to debugging applications. See Chapter 12 for more information on software interrupts.

11.6 Real Time Multi-Tasking

Let us take time to review our progress to this point. First we covered the basic structure of programming: in-line code, loops, and subroutines. These simple structures work well for some jobs, but they lack flexibility and the ability to rapidly respond to an external event. This is due to the fact that these simple programs can only do one thing at a time. They must complete each task before they can do anything else.

Interrupts, as we have seen, are an improvement over the basic programming structures. Interrupts allow the system to rapidly respond to an external event. This is accomplished by suspending the task they are working on and immediately executing the interrupt task. When the interrupt task is done, control is returned to the interrupted task. Interrupts essentially allow a program to respond asynchronously to events, while noninterrupt driven programs must respond to events synchronously.

This brings us to the definition of *real time*. Like many technical terms, this one means different things to different people. In general, the term is used to describe a system that must respond to an event in a fixed and generally short period of time. Examples of real time systems include engine control computers and process equipment.

Interrupts are a powerful tool for building real time systems. For many simple applications an interrupt driven architecture is sufficient. Even an interrupt driven program, however, has a fixed architecture. Each task is executed in sequence and its place in the sequence is fixed. True, we can choose to execute or not execute a task by using conditional jumps, but we cannot dynamically alter the basic program architecture.

The simple program architectures previously discussed have worked well over the years and are still the mainstay of most microcontroller programming. As newer and more powerful microcontrollers come on-line, however, they must be programmed with more powerful architectures. Consider, for example, some of the features of Motorola's 68HC11's resources:

- Two serial channels, one synchronous and one asynchronous.
- Two bidirectional I/O ports.
- One input-only port.
- One output-only port.
- Eight channels of analog input.
- Five pulse width modulated outputs.
- Watchdog timer.
- Real time clock.
- External and internal interrupts.
- Three timers.

This is a large number of channels over which data can be coming and going. Coming up with a single program architecture that can efficiently utilize all of these resources is nearly impossible.

What is needed is a way of dynamically assigning tasks as they need to be done. For example, consider a situation in which our microcontroller is connected to a remote terminal. The operator types in a command one character at a time. A classic program loop would check on each pass through the loop to see if a character had been received. Even if the program is interrupt driven, it would still need to make such a check because all the interrupt routine does is allow us to rapidly respond to the character arriving.

More processing responsibility could be put on the interrupt routine, but this is a bad idea. The whole point of an interrupt is to allow the system to respond rapidly. If a large part of processing time is spent on one interrupt routine, the system cannot be servicing other interrupts.

Let us consider this scenario as an alternative. As previously, Our interrupt routine simply gets the incoming character and saves it in memory. If the character is a carriage return or some other character signifying the end of a command, the interrupt routine will dispatch a "command processing task." This command processing task can in turn dispatch other tasks as necessary. And each of these tasks can also dispatch tasks.

With a little visualization, one can get a feeling for how this works. Each task spawns other tasks as necessary for the job to get done. When the job is done, tasks stop spawning. Or in some situations, tasks may spawn themselves over and over as needed. These tasks are called *self-posting*.

Figure 11-8 shows conceptual models for a conventional program and for a multi-tasking system. Notice that the program in Figure 11-8a is a complete, stand-alone entity. In Figure 11-8b there are a variety of tasks, all coordinated by the SCHEDULER. We will look at the SCHEDULER more closely later, but for now it is enough to understand that the SCHEDULER simply executes tasks as interrupts or other tasks schedule them.

In such a system it is impossible, of course, to say ahead of time exactly what tasks will be executed or when. While this is somewhat of an intimidating thought (shades of HAL in *2001: A Space Odyssey*!), it has some major advantages.

The program flow is dictated by circumstances. If a lot of serial I/O is happening, many serial tasks get dispatched. If a lot of analog inputs are coming in, many analog tasks get dispatched. The flow dynamically adapts itself to the circumstances. This is a much more elegant way of dealing with real world situations than trying to come up with one fixed architecture that can handle any situation.

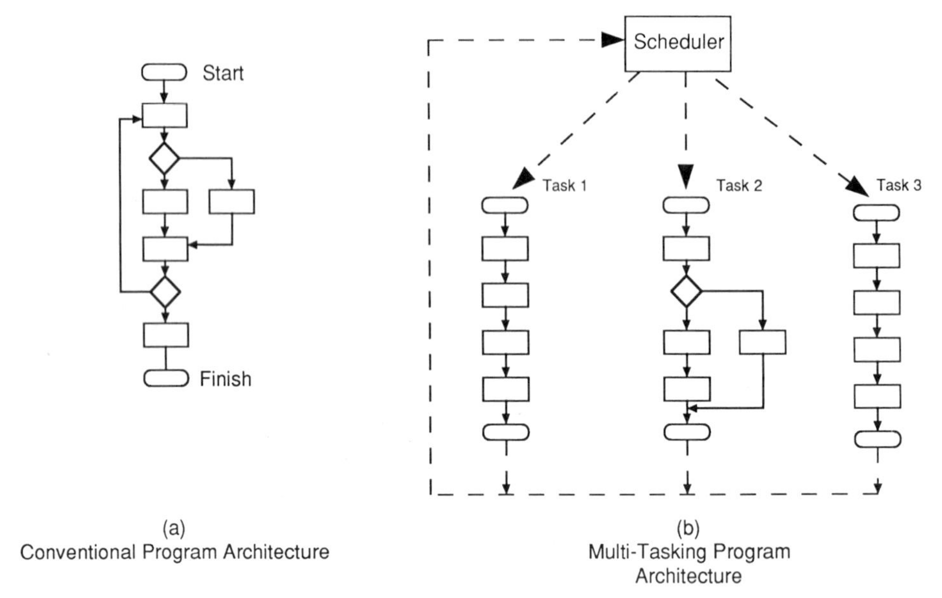

Figure 11-8
Conventional versus Multi-Tasking Program Architectures.

Now, the only real question is "how do we handle all of this?" We could simply let each task call, via a JUMP instruction, the task that it wished to invoke. In fact, this is how high-level "threaded" languages work. For our purposes, however, it is better to build a table of the tasks that are to be executed. There are a number of advantages to this approach, some of which will be covered later.

One advantage that should be mentioned now is that a table allows one task to schedule several other tasks. Going back to the remote terminal example, a task may wish to ask for an input such as a target temperature from the operator. This process would normally be broken down into two tasks:

```
POST DISPLAY_TPROMPT ; Prompt for the temperature.
POST GET_COMMAND     ; Get a command from the operator.
```

Our task could now terminate or simply suspend itself until the command was entered by the operator.

The actual operation of the POST directive depends upon the microcontroller and the choice of the programmer. POST could simply be a JSR instruction or it could be implemented with software interrupts.

Now to develop some of the mechanics of the multi-tasking structure. A typical task table would look like the one in Figure 11-9. The table

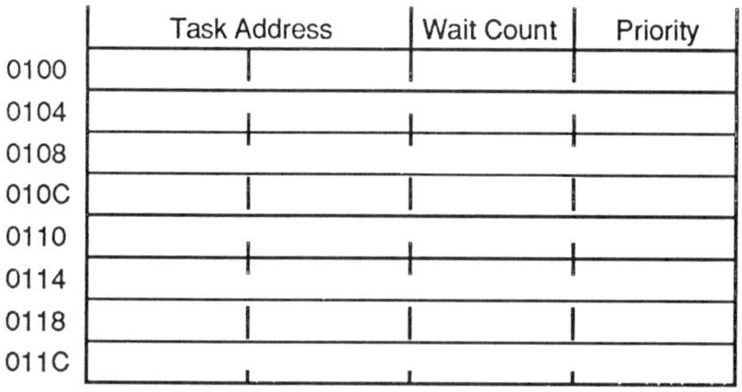

	Task Address	Wait Count	Priority
0100			
0104			
0108			
010C			
0110			
0114			
0118			
011C			

Figure 11-9
Task Table.

uses four bytes for each entry. The first two bytes hold the address of the routine to be executed; the third byte holds a sleep count; and the fourth byte controls the priority of the task. The last two items will be discussed in a few moments.

Some resources are needed to handle the table. One is the POST routine for putting entries in the table and another is a SCHEDULER. The SCHEDULER's job is to pass control to the tasks in the table. When the task is done, it returns control to the SCHEDULER.

The SCHEDULER also needs a real time interrupt routine to handle the timing. These resources will be examined in detail but first some background is necessary.

Entries will continuously be made in the task table. As each task is completed, its entry will be deleted. Since entries can be POSTed by interrupts or other routines, there is no way of guaranteeing that the various tasks will be executed in the order they are POSTed. In our prompting example, the task to display the prompt could have been the last entry in the table. The POST routine would then put the task entry for the GET_COMMAND task in the first available slot in the table—ahead of the display task.

The net effect of this would be that the operator would never see the prompt since the system would be waiting for a reply *before* it prompted.

There are a number of ways of overcoming this *task synchronization problem*. One of the simplest is to use the priority bits in the task table. We could assign the GET COMMAND task a relatively low priority, say 4. We could then assign the display task a relatively high priority, say 5. The SCHEDULER will dispatch the highest priority tasks first, so the display routine will always be executed first and the GET COMMAND task second.

Another useful feature that often comes up is the ability to schedule tasks. This is what the WAIT field was reserved for in the task table. The way this works is as follows: Periodically, say every 50 msec, a timer generates an interrupt. The interrupt routine simply goes through the task table and decrements the WAIT count for each task. If the SCHEDULER is built so that it will only execute tasks with a WAIT value of zero, then it will provide a handy timing function.

This real time scheduling accomplishes two things. First, it provides another means of synchronizing tasks. For example, the prompting sequence just discussed could have POSTed DISPLAY_TPROMPT with 0 delay. GET_COMMAND could then have been POSTed with some reasonable delay, say 100 msec. In this case, delaying a task is not as good a means of synchronizing as is the use of priorities. We must guess how long it will take to complete the display task. But for some applications, delaying is quite useful.

Second, delaying a task can also help make better use of the CPU's bandwidth. For example, in many control applications there is a long period between the initiating of an action and the actual consequences.

A good example is the fail-safe on some mechanical relays. If it is critical that a relay should reliably close or open, an extra set of contacts are often specified. The purpose of the extra contacts is to signal back to the controller that the appropriate action has been taken.

An example is shown in Figure 11-10. The relay can be checked for positive closure by reading the input on PA1. If the relay has successfully closed, PA1 will read as a 0, thus insuring that power has at least been applied to the motor.

Mechanical activities can take a long time with respect to a microcontroller's time frame. Our relay might not close for 50 or even 100 msec. A microcontroller can get a lot of work done in this time, so it

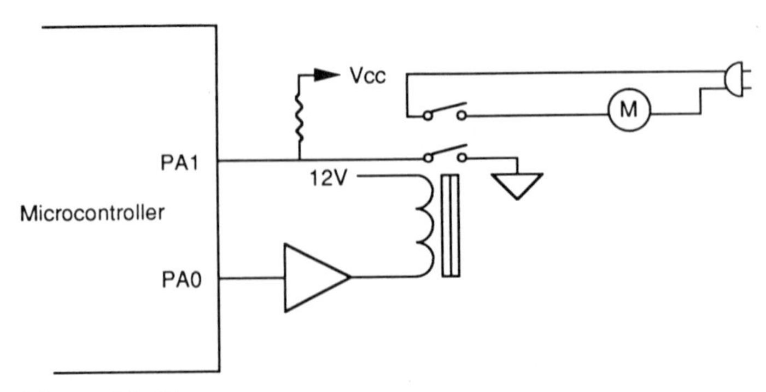

Figure 11-10
Fail-Safe Monitoring.

would be silly to just keep polling the input to see if the relay closed. A better solution is to use two tasks and delay one of them:

```
POST MOTOR_ON,0        ; Turn the motor on ASAP.
POST CHECK_MOTOR,100 ; In a 1/10 of second, check it.
```

This allows the microcontroller to set the relay immediately. It can then accomplish other tasks until the 100 msec waiting interval has elapsed. Only then does it spend processor time on checking to see if the relay actually did close.

Now that we see what we want, how can such a system be achieved? The following discussion will present descriptions of a software architecture that will implement these features. Actual code is not presented since the code will vary from microcontroller to microcontroller. The algorithms and flowcharts *will* be presented. From these, the actual coding is straightforward for whatever target microcontroller you will be using.

In the following example, an empty position in the task table is indicated when the address field contains a value of 0000H. The first algorithm presented is the POST routine. The flowchart for the POST routine is shown in Figure 11-11. POST's operation is quite simple: It searches for the first empty slot in the task table and then copies the address of the task to be executed, the task's priority, and the sleep count into the task table. When it has done these operations, it simply returns control to the calling routine.

As noted previously, multi-tasking systems are dynamic; the POST routine must be prepared to deal with the possibility that the task table may be full. If this happens, some type of error handler must be invoked. What the error handler actually does is up to the individual system, but there are several standard options. The first and simplest is to return without doing anything. The task will never be posted and thus will never be executed. This may or may not be acceptable, depending upon the application.

Another option is to invoke a software reset sequence. This is tantamount to pushing the panic button. In many applications, however, a clean reset is preferable to an uncontrolled crash.

A third possibility, and generally the most preferable, is for the error handler to simply keep executing the POST task until a slot is freed up in the table. This can be complicated, however, since several tasks may be trying to execute the POST routine.

All three of these options have advantages and disadvantages. The specific application must dictate which is chosen. The important thing is that some means be available to *predictably* handle system overflows.

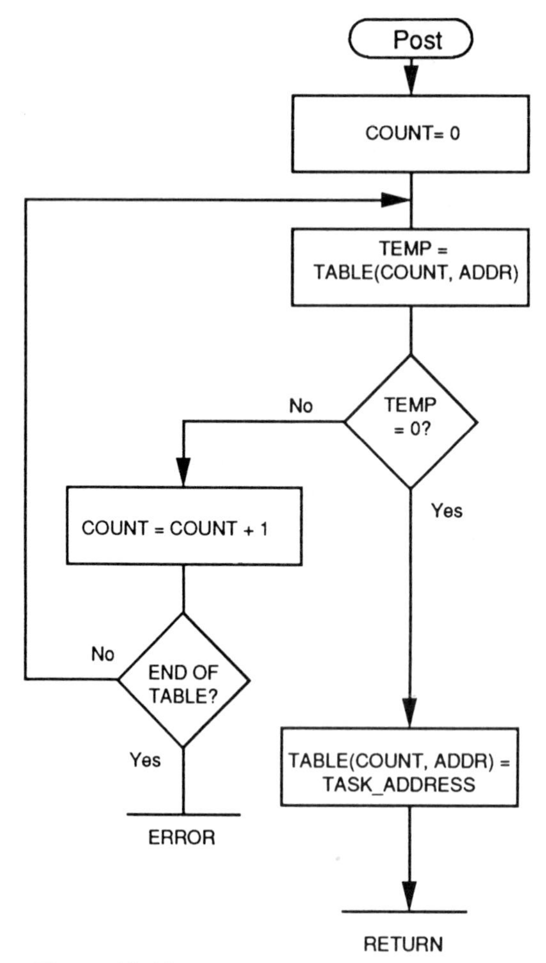

Figure 11-11
POST Routine.

Now that the task table has been defined and we have a way of getting tasks into it, let us look at how the tasks get executed. Figure 11-12 shows the flowchart for the SCHEDULER routine, which is the heart of the multi-tasking executive.

The first thing the SCHEDULER does is to determine the highest priority of any tasks in the task table. There are several ways to do this. One of the most direct ways is to simply scan the table looking for the largest value.

Once the highest priority of the tasks is known, the SCHEDULER can begin searching the table for tasks to execute by simply stepping

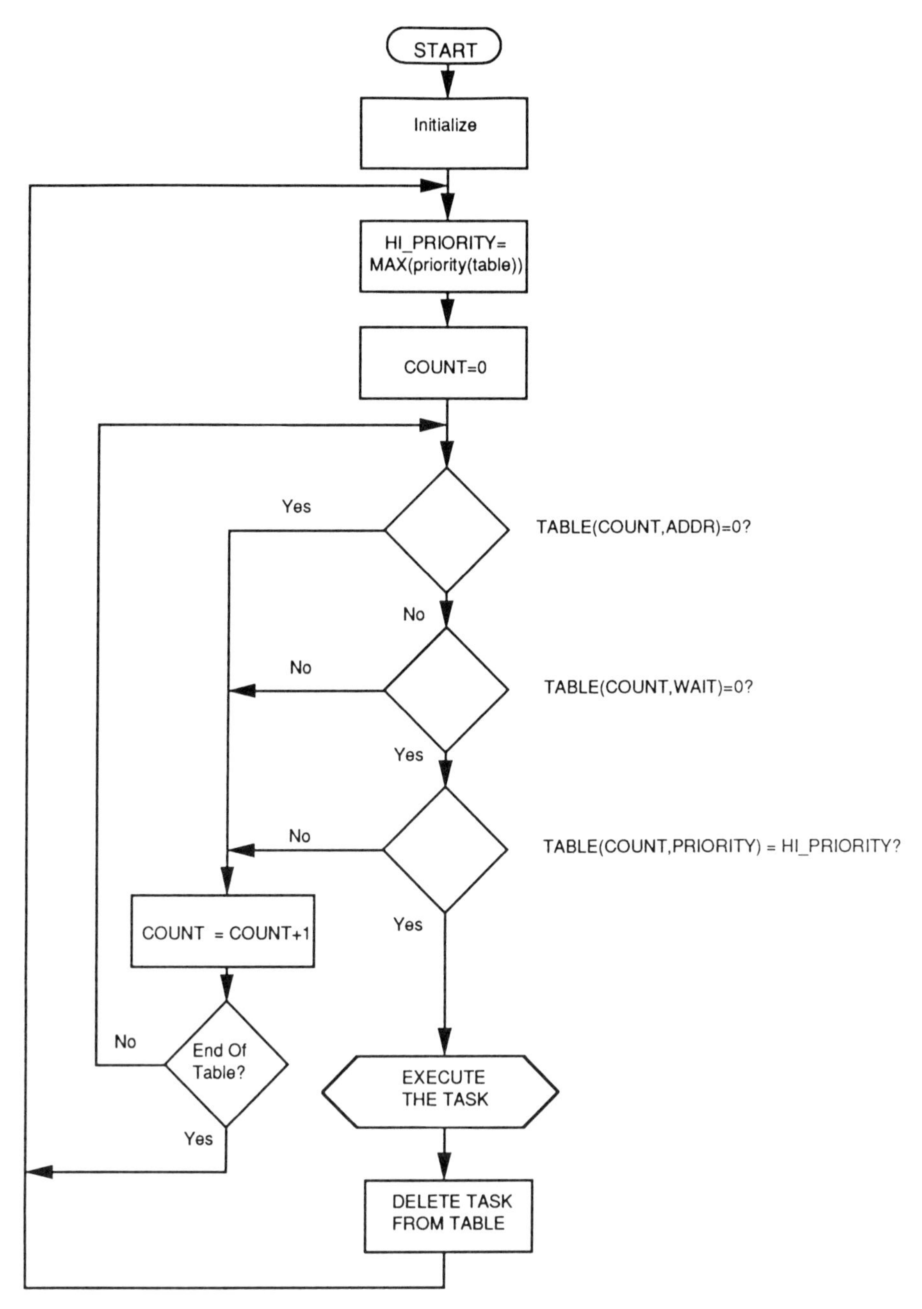

Figure 11-12
Scheduler Routine.

down the table looking for tasks. It first checks to see if the task is ready to execute. This is indicated by a WAIT count of zero for the task. Next, the SCHEDULER checks that the task's priority is equal to what was determined to be the highest priority.

When all the above conditions are met, the SCHEDULER will invoke the task. This can be done in a variety of ways. A software interrupt sequence can be set up or the SCHEDULER may simply do an indirect JUMP to the address stored in the table. Once the task has completed execution, it returns control to the SCHEDULER. The SCHEDULER will then delete the task from the table by simply writing a zero into the address portion of the table. The SCHEDULER then loops through this sequence continuously.

In the above discussion, we pointed out that the execution of a task can be delayed by POSTing it with a WAIT value. The SCHEDULER does not execute any routine until the WAIT value is zero; so the obvious question is "how does the WAIT value get to zero?"

The answer is in the final routine of our multi-tasking executive: the TICK routine. The flow of the TICK routine is shown in Figure 11-13. The TICK routine is a periodic interrupt routine. Typically, it will be set up to occur ever 10 to 50 msec. What it does is to simply step through the task table, decrementing the count each time as it goes. As a practical matter, if the WAIT count is already zero, the variable is not decremented.

The TICK routine is what sets the overall pace of the system. It should occur fast enough that the system has good response yet slow enough that tasks can make practical use of WAIT values.

The preceding discussion has been, of necessity, fairly high level. Like many concepts, multi-tasking may seem somewhat complex at first pass. However, once you have actually coded the above algorithms for a particular machine, the concepts are easily understood.

Let us look at another typical application: an engine control computer. Figure 11-14 shows the basic layout of a microcontroller used to implement an electronic ignition system. This is a very common use of microcontrollers in automotive applications.

Before we discuss the operation of the circuit, it is worthwhile to briefly review the principles of the internal combustion engine with fuel injection. Let us assume the piston is at the very top of its stroke. This is called "top dead center" or TDC. As the piston moves down, the intake valve opens and air is sucked into the cylinder. As the piston comes back up, the intake valve is closed, the fuel injector is turned on, and a squirt of fuel is injected into the cylinder. When the piston reaches TDC, the spark plug is fired. This causes the fuel/air mixture to burn. The burning gases force the piston down. As the piston comes back up again, the exhaust valve is opened and the burnt gases are pushed out of the piston. The

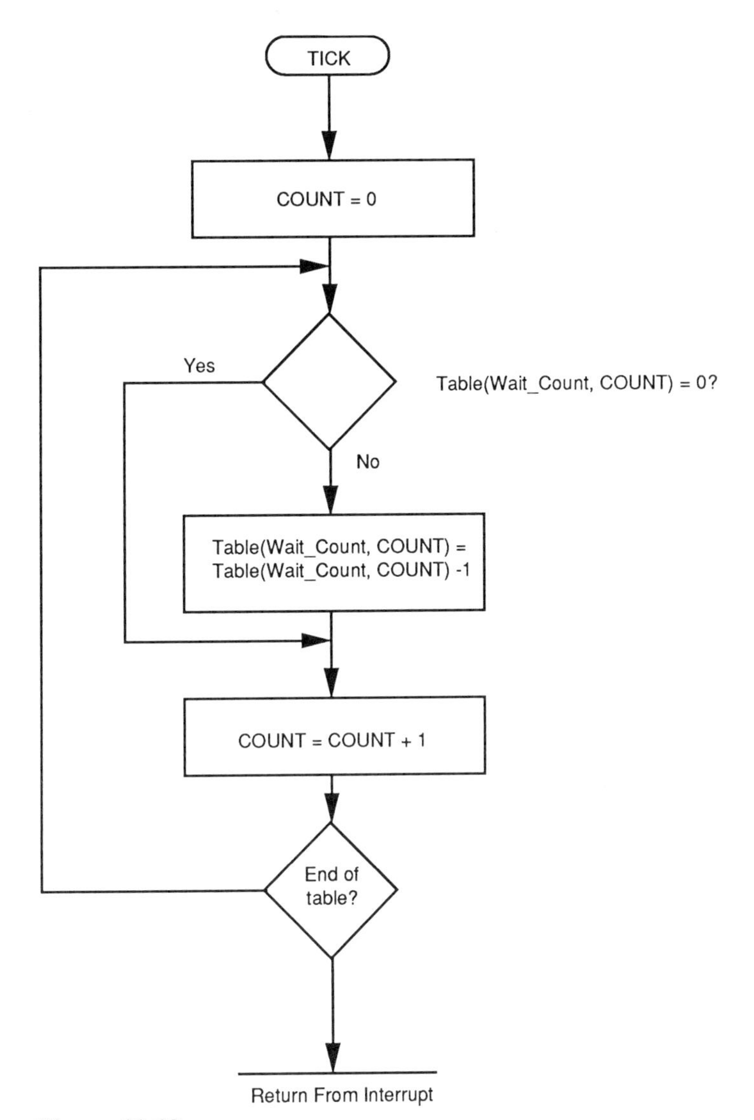

Figure 11-13
Periodic Interrupt TICK Routine.

cycle then starts over. These four operations (intake, compression, combustion, and exhaust) form the basic operation of the common four-cycle internal combustion engine.

For this scenario to work properly, the timing must be precise. The opening and closing of the valves is handled by mechanical linkages. The firing of the spark and the injecting of the fuel are done electronically.

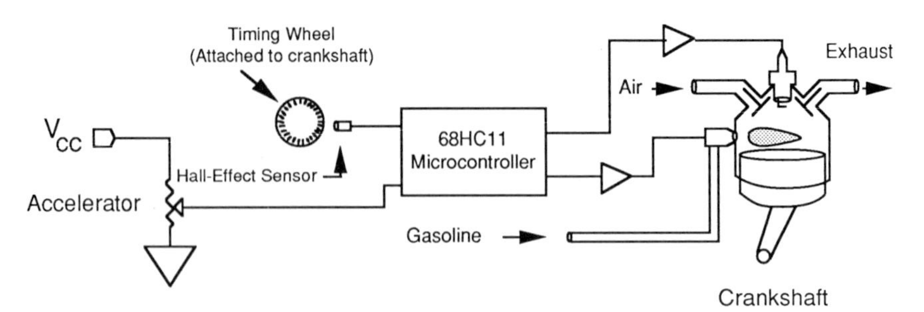

Figure 11-14
Simplified Diagram of an Engine Control System.

There are several key things that must be taken into account in the timing. First, the fuel injectors must be turned on at the correct time and for the correct length of time. The longer the injectors are on, the more fuel that will be injected into the cylinder. The more fuel that is burned, the faster the piston will move and thus the higher the engine RPM.

A second key factor is when to fire the spark plug. The idea is to have the fuel burning just as the piston begins its downstroke. This produces a smooth, efficient pulse of power. If the engine is running slowly enough, the spark plug can be fired exactly at TDC.

In practice, however, there are several delays in the firing sequence. It takes a certain amount of time for the spark to form. It also takes a finite amount of time for the fuel/air mixture to start burning. As the engine RPM increases, these delays become significant. To compensate for the delays, the spark plug is fired in *advance* of the piston reaching TDC. This is what is meant when a mechanic speaks of adjusting the "advance" of the engine timing.

The microcontroller knows exactly where the engine is in the cycle by monitoring a special wheel attached to the crankshaft. The wheel has alternating magnetic domains (called timing marks) that activate a Hall-effect sensor as the wheel moves. The driver sets the amount of fuel the engine gets by pushing on the accelerator pedal.

The accelerator pedal is connected to a potentiometer. As the pedal is depressed, the voltage moves closer to ground. This analog voltage is fed into the analog-to-digital converter (ADC) on-board the 68HC11. The 68HC11 then calculates the length of time the fuel injectors should be turned on for any given acceleration. This sequence is not particularly time sensitive since the mechanical effects of the engine's inertia set a minimum time in which the engine can perform.

On the other hand, the timing of the spark and the activation of the fuel injector are quite sensitive. If we tried to monitor the engine position by polling the Hall-effect sensor, we could get very erratic operation.

The main program must spend a relatively large amount of time monitoring the ADC, calculating pulse widths and timing information, and other such chores. By the time the program got back to the point in the loop where it could read the Hall-effect sensor, the proper timing mark would have long passed.

To solve this problem, the Hall-effect sensor is connected to the interrupt line. When a timing mark is reached, the microcontroller will drop whatever it is doing and immediately respond to the interrupt. The interrupt routine will use the 68HC11's timers to schedule the correct time for the spark to be fired and the correct time for the fuel injectors to be turned on.

This is a simplified example, of course. In practice, the engine control computer must monitor the intake air flow, intake air temperature, oxygen level, battery voltage, and other such data. Aside from the spark and the fuel injector, the controller will also typically be controlling the solenoid for the EGR valve, a stepping motor for idle control, cruise control settings, etc.

As complex as this application is, it is nevertheless quite well defined. As such, it would probably be better implemented as a straight program loop with critical events coordinated by the interrupt structure. On the other hand, many of the tasks, such as turning on the fuel injectors for a specified time, lend themselves well to the multi-tasking architecture. As the processing power of microcontrollers increases, multi-tasking undoubtedly will become a common architecture for such real time applications.

Many of the 8 bit controllers and virtually all of the 16 or 32 bit controllers are good candidates for multi-tasking executives, at least in some applications. One of the best is Motorola's 68HC11, since it is designed with this type of operation in mind. The simpler 6801 can also be used, as can Intel's 8051 and 8096 family of devices. Simpler eight bit units such as the 6805, 8748, or most of the four bit units are not powerful enough to utilize multi-tasking.

Unlike general purpose computers, microcontrollers are not used in "multi-user" environments. Therefore, many of the requirements normally associated with multi-tasking and multi-user operating systems are not present. For example, microcontrollers do not run in *protected* or *system* modes, and they do not have address exception registers to prevent one task from stepping on another task.

In the more self-contained world of the microcontroller, these are not generally major concerns. Programs are normally stored in ROM, not loaded into RAM for execution. Thus it is unlikely that the program will be accidentally overwritten by some other routine. The program debugging is done in the development environment, so the possibility of the user generating code that could go awry in the field is limited.

Finally, we present some general comments on using multi-tasking executives. Multi-tasking is inherently modular. The process of defining and coding the tasks tends to lend modularity to the program. This modularity will generally follow through into the development and debugging effort.

Modularity is extremely valuable when going through the fault isolation process. Problems are more easily localized to a certain task or at least a group of tasks. This is a fortunate thing, since the dynamic nature of a multi-tasking system can make problems appear and disappear at intermittent intervals, which can be very frustrating.

An investment in adding diagnostic and "debug" features into the multi-tasking environment can pay off handsomely. One typical feature is task tracing. If in the debug mode, the SCHEDULER lists each of the tasks as it is called. If the development system is connected to a terminal or a PC, the task list can be printed out. This provides a visual record of exactly what steps were taken during program execution.

Another useful feature is a switch that allows the SCHEDULER to dispatch routines one at a time. In this way the execution of the program can be slowed down to observe functions on a step-by-step basis.

Some simple things can also be done. For example, the area of the memory occupied by the stack can be cleared to all zeroes. After a certain period of time the stack area can then be displayed. This gives a good idea of what percentage of available RAM the stack is using. Stack overflows are a common but difficult to identify cause of many software failures. Some mechanisms for monitoring the percentage of stack usage can be very valuable.

A very popular option is the use of a *watchdog timer*. Many microcontrollers come with these built in or they can be added externally without too many parts. The idea behind the watchdog timer is simple: if the timer ever times out, the microcontroller is reset. It is the job of the program to reset the timer *before* it times out. For example, the SCHEDULER could reset the watchdog timer at the start of each loop. If any task were dispatched that did not complete in a reasonable amount of time, the watchdog timer would time out and a system reset would be generated. This scheme also protects against power brownouts, noise spikes, or anything else that may lead to a program going out of control.

Multi-tasking systems are fun, and the improved bandwidth they provide for microcontrollers insure that they will become a more common part of the software architecture of future products.

11.7 Chapter Summary

Software for microcontrollers is much more algorithmically oriented than the software for microsequencers or PLDs.

- Microcontrollers are typically programmed in assembly language and rarely in high-level languages.
- The *source file* contains the human readable *mnemonic* form of the machine instructions.
- The source file is assembled to form the *object code* file. The object code is composed of the actual *machine instructions*.
- When writing programs in either assembly language or a high-level language, it is important to *structure* the program: Make the inputs and outputs of a routine clearly defined; keep routines to manageable size; and connect several simple routines together to accomplish complex tasks.
- *Interrupts* add considerable power to the processor. They allow the processor to respond to events as they occur.
- *Multi-tasking* is generally the best way to make optimum use of a processors bandwidth in complex applications. Multi-tasking systems break the overall job down into a series of tasks. The tasks are invoked as needed to get the job done.

12

Additional Tools of the Trade

Understanding the architecture and theory of programmable circuit design is only part of the knowledge base needed by the modern designer. Equally important, and in many cases more important, is an understanding of the tools available for developing programmable products.

So far we have looked at some of the basic software development tools. These include PLD assemblers and compilers, microcode assemblers, and conventional object level assemblers. These are, of course, the mandatory tools for configuring or programming the physical IC. However, just programming the device is not enough. The programmed part must be tested and evaluated in the actual product. This chapter presents a variety of tools designed to help get a working product out the door.

12.1 Basic Tools

The fundamental tools necessary are the ones commonly found around the typical electronics lab. These include voltage meters, logic probes, and the simple but ever valuable set of clip leads. While these basic tools are common, the importance of having them available should not be underrated. There is a tendency on the part of many engineers, particularly when sitting down at the bench to debug a sophisticated FPGA, to assume that complex tools and procedures are required.

In practice, no debugging work should begin until some simple initial steps have been taken. First, what are the symptoms? Usually during the development phase the answer is likely to be a vague, "It doesn't work!" Nevertheless, it is a good idea to collect and mentally integrate all available information about the problem. Then a quick mental hypothesis as to what the problem is can be formulated. Only when this is done should the first diagnostic instrument be picked up. It is truly amazing the amount of time that engineers and technicians waste by simply probing around a board in the forlorn hope of tripping across the problem.

The first diagnostic test should *always* be a power check. This includes those cases where everybody *knows* the power is good. It only takes a second to see that the V_{cc} rail is where it is supposed to be, and it can save hours of fruitless probing.

The next step should always be to test the static lines. A logic probe or a DVM is handy for this. Are the reset lines in the correct state? Are any static enables at the correct potential? This is particularly important when using the bidirectional pins on PLDs. A common problem is to find that a pin is inadvertently in the input mode when it should be an output or vice versa. Simulation should catch this type of problem but often does not.

After checking the static conditions, a quick check should be made on the clock lines. This is where a good oscilloscope comes in handy. Are the clock signals reasonably clean? Are the levels correct? Check *both* the AC and DC line levels. An otherwise fine looking clock signal can have a DC component on it that puts the signal outside of the threshold limits.

These simple checks will catch a surprisingly large number of problems. And since they are easily and rapidly performed it is a good idea to go through them periodically. Even if the engineer simply shuts down to go to lunch, it is a good idea to run these tests upon return. First, this will help catch intermittent problems like oscillators failing to start correctly. And second, it minimizes the impact of test equipment being borrowed and returned but either not hooked up correctly or not hooked up at all.

When everything looks OK with the simple stuff, it is time to bring in the big guns.

12.2 Logic Analyzers

The digital engineer's strongest tool is the logic analyzer. Functionally, the logic analyzer is a specialized computer with a very wide and very fast parallel input port.

The logic analyzer can format the data from the input port in a variety of ways. Waveforms or bit tables are the most common, but sophisticated logic analyzers can even disassemble object code and display the acquired data as assembly language mnemonics.

Functionally, the logic analyzer is connected to the key lines of the circuit. For example, when debugging an address-decoding PLD, the logic analyzer would be connected to the high-order address bits going into the PLD. The output of the PLD would also be monitored by the logic analyzer. By comparing the address data on the input with the selected outputs, the PLD's proper operation can be verified.

In many applications, the large volume of digital data would quickly swamp even the largest logic analyzer's memory. To minimize the amount of data acquired and to speed up the process of sorting through the data, logic analyzers can be set to trigger on specific events.

For example, Figure 12-1 shows a typical two-level address decoder using PLDs. PLD1 decodes the high-order address lines. When these lines have the address of PLD2 on them, the PLD2EN line goes active. PLD2 then further decodes the low-order addressing.

For discussion purposes, let us assume that during system checkout it is observed that there is a problem with the devices connected to the outputs of PLD2. The devices are not responding correctly when addressed. Let us assume that PLD1 has checked out as correct. The objective is to then determine if PLD2 is functioning correctly.

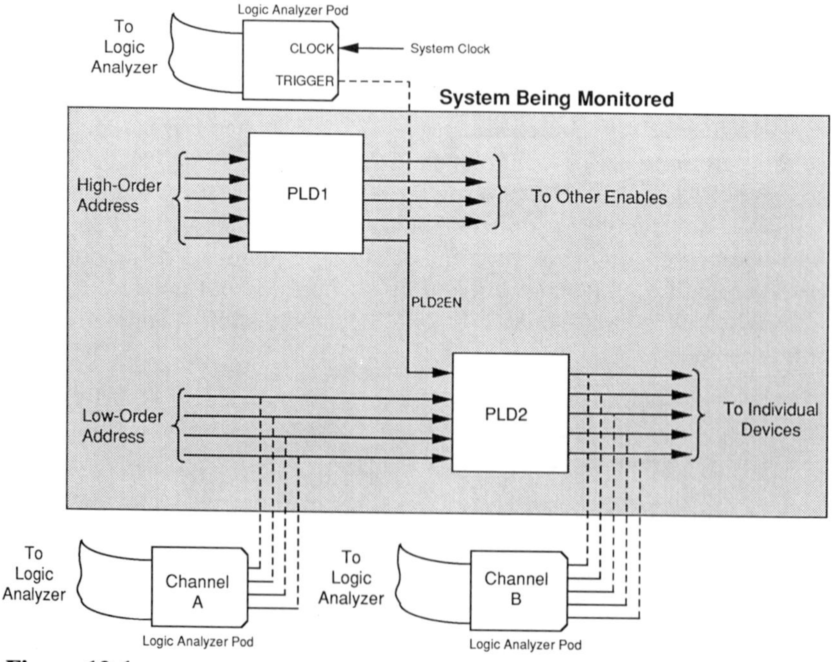

Figure 12-1
Typical Logic Analyzer Setup.

All of the inputs and outputs of PLD2 could be monitored asynchronously. However, there may be a great deal of bus traffic before PLD2 is first exercised. This would overflow the logic analyzer's buffers before the data of interest could be collected. The solution is to connect the trigger input of the logic analyzer to the PLD2EN signal. The analyzer will only collect data when PLD2 is being addressed.

When PLD2EN goes active, the data is collected and stored for display. The display will only include the bus traffic for the periods when PLD2 was selected. This makes it an easy matter to see if the outputs responded correctly for the given inputs.

It is interesting to note that the trigger does not necessarily control the data *acquisition*. For example, the analyzer may simply store all data coming in for as long as its memory allows. When the memory fills up, the oldest data is discarded and the new data is saved. This allows triggers to be set *before* or *after* an event occurs as well as *when* it occurs. Triggering on an event before it happens does violate one's intuitive perspective on time. However, the technique can be very valuable when debugging sequential circuits such as state machines. The problem condition can be set as the trigger event. The analyzer will then show what conditions led up to the problem.

Logic analyzers typically have a large range of options for setting these trigger events. The ease of use and flexibility of the trigger events is one of the main points to consider when selecting a logic analyzer.

For simple PLDs such as PROMs, PLAs, and PALs, the logic analyzer may be the only debugging tool available. A good analyzer and the skill and experience to use it are vital in developing products using PLDs.

12.3 Monitors

The use of monitors is restricted to microcontroller-based systems. Monitors are special programs designed to help in code development. Most monitors are designed to interface a microcontroller via its asynchronous serial port to a stand-alone computer. The stand-alone computer acts as an interactive terminal and code development station for the microcontroller.

The purpose of the monitor is to allow the engineer to communicate with and control the microcontroller. The monitor program typically provides the following functions:

- The ability to download a program from the development station to the microcontroller.
- Once the program is downloaded, the monitor provides a means to

execute the program. Either an EXECUTE or a CALL command can be used. These commands are described later.
- Functions for examining and changing the values stored in the microcontroller's memory and registers.
- Single-step capability allowing the program to be executed one step at a time. Usually the contents of the registers are displayed after each step.
- Instruction tracing, which is similar to single stepping except that a specified number of instructions are executed.
- Break Points which, as the name implies, are places where the program will stop executing and return to the monitor. Break points are handy for checking to see if a section of code is ever actually executed.
- Many monitors provide a one-line assembler. These are handy for patching a program instruction. This eliminates the need to reassemble the program and download it again.

It is worthwhile to examine how these functions are implemented. There are two reasons for this: First, understanding how the monitor works provides some interesting insights into microcontroller software. Second, it is important to understand exactly what the monitor is doing. This impacts the way the code is written and debugged.

Most microcontroller assemblers output the object code as an ASCII text file. This file will generally be in one of several standard formats. Motorola's "S" format and Intel's Hex format are the most common. Each string in the file is called a *record* and will contain some specific information:

- The type of record (executable code, debug information, etc.).
- The address at which the first byte of data is to be stored. All of the other bytes are stored at incremental addresses.
- The number of bytes to transfer.
- The ASCII encoded object code and data for the program.
- A checksum, which is normally the 2's complement arithmetic addition of the individual characters in the string. When the characters in the string and the checksum are added together, the result should be zero.

The purpose of the checksum is to insure that the data is transmitted properly. A checksum other than zero indicates that noise or some other problem has corrupted the transmitted data.

Like the source file, this output file will be resident on the development station. To be of any use, the code must be transferred to the microcontroller's program store. This process is called *downloading*.

Normally, downloading is done via a serial link connecting the development station and the microcontroller. A communications program on the development station sends the formatted information from the ASCII file, and the monitor program running on the microcontroller receives the data.

When the monitor is commanded to receive a download program, it converts the ASCII address information and sets its pointer to that address. Each of the two ASCII characters received are then converted to binary and stored as a single byte in memory. A running total of the characters is kept. When the last data character of the record is received, the running total is added to the checksum. If the total is not zero, then an error in transmission has occurred and the data must be re-sent.

The EXECUTE command in the monitor simply does a JMP (jump) to the address at which the program is stored. All control is therefore passed to the download program. The CALL command, on the other hand, does a jump to subroutine (JSR) to the program address. If the program ends with a return from subroutine (RTS) instruction, the monitor will resume executing when the program terminates.

Downloading and executing the program are, of course, handy functions. It can be difficult, however, to find problems in a program when it is running at full speed. This is where the other features of the monitor come in.

For example, we may suspect that a particular branch of the program is not executing quite properly. An easy way to check this is to set a break point at the top of the branch. When the conditions are reached in the program that cause the branch to be taken, the break point will be triggered. Control will be transferred back to the monitor. At this point, the branch can be single stepped through to see if a problem exists.

Most monitors make use of the software interrupt for this purpose. The mechanics of this are:

- The monitor saves the value of the instruction at the address of the break point.
- A software interrupt (SWI) instruction is then stored at the break point address.
- The vector for the SWI points to a special interrupt routine in the monitor.
- When the program reaches the SWI instruction, the contents of the registers are pushed on the stack.
- The SWI routine is then invoked. The routine will replace the SWI instruction at the break point with the original instruction.
- Since the values of the registers at the time of the interrupt are on the stack, these values can be displayed. It is important to remember

that the monitor displays the values *at the time of the interrupt.* It is common (and confusing) for this display to be called the *current* register status. The monitor is using the registers for other things by the time the status information is actually displayed, so it is not really the state of registers being displayed.
- The monitor's command processor is then invoked.

There are a couple of implications to this operation. First, to work properly the break point must be set at the first address of an instruction. Otherwise the SWI instruction will not be recognized and the results will be unpredictable. The second implication is that break points can only be set in RAM or EEPROM program stores. Break points cannot be set in programs stored in ROM or EPROM using the SWI technique.

Single-step operation can be implemented in several ways. It may use exactly the same sequence as the break point. Or, to allow the single stepping of ROM and EPROM routines, a hardware interrupt may be used. A typical single-step sequence would go something like this:

- A timer is set to generate an interrupt. The delay is set just long enough to allow a JMP instruction to the single-step instruction address to be executed.
- As soon as the target instruction begins to execute, the interrupt becomes active. The processor will complete the instruction before executing the interrupt.
- The rest of the sequence is the same as with a break point.

The overall purpose of all of these debug features is to provide visibility into the execution of the program being developed. The better the visibility into what the program is doing, the easier it is to find and correct problem areas.

Often, the monitor makes a good starting point for code development. In fact many products may simply extend the monitor until it becomes the final product. This makes production testing and field maintenance much simpler.

Some microcontrollers are sold with monitors built into their ROMs. The Motorola 6801L1 discussed in Chapter 10 is a classic example. Other monitors are available for most microcontrollers. These are often supplied by the microcontroller vendor or can be purchased from third-party vendors. Motorola, for example, maintains a free access bulletin board that has monitors and other support products available.

In most applications, the microcontroller must be used in an expanded mode with additional RAM and EPROM. This of course uses up the available I/O ports, so the lost ports are often emulated with discrete logic.

In some way or another, all monitors take up some of the microcontrollers resources. These lost resources must be compensated for, either directly or indirectly. One common approach is to design a development version of the circuit that is similar to the intended product but has the extra RAM and EPROM. The code is developed and debugged using the monitor on the development station. Once the program has been written and debugged, it can be transitioned to the real product. Obviously, the closer the development version is to the real product, the smoother the transition will be.

Finally, to make use of a monitor requires a fairly sophisticated microcontroller. Many of the simpler microcontrollers do not have sufficient hardware or software resources to make use of a monitor program. Developing programs for these products requires other tools such as the ones to be discussed next.

12.4 Simulators

Simulators make use of the superior capability of a general purpose computer to *model* a circuit. The simulator is simply a program that runs on the computer. There are a variety of simulators available. The various types can be divided into three basic classes:

1. "Built-in" Simulators. These are generally functional simulators included in the PLD development packages.
2. General Purpose Simulators. These can be further broken down into analog, digital, and mixed signal simulators.
3. Device Specific Simulators. These are usually confined to simulation of the resources of microcontrollers.

12.4.1 Built-in Simulators

Built-in simulators were discussed in Chapters 5 and 6. To briefly review, built-in simulators for most simple PLDs are designed to provide functional validation of the designs. Timing and other parameters are *not* simulated.

Some of the more specialized tools available for the more advanced programmable gate arrays, on the other hand, do provide timing and other information. However, all of these simulators are designed to simulate the PLD at the device level. To get a simulation at the board or system level, a general purpose simulator is required.

12.4.2 General Purpose Simulators

As the name implies, general purpose simulators are designed to simulate conventional circuits at the gate or board level.

Analog simulators are designed to solve the matrix of differential equations that describe analog circuits. This type of analysis is not very informative for digital signals, so the engineer working with programmable logic normally will make little use of purely analog simulators. Occasionally, however, these tools can be useful for studying specific phenomena such as metastability.

Digital simulators abound and take a variety of forms. Most simulation is done at the chip level and as such is not normally of direct interest to most design engineers. However, since programmable logic designs are often migrated to gate arrays or semicustom designs, the engineer working with programmable logic generally gets some exposure to these tools.

The quality and ease of use of the chip-level simulators varies. In general, these simulators are similar to the ones discussed for programmable gate arrays. The simulators insure that timing margins are met, that clock skew is not a problem, and that the overall functionality of the circuit is correct.

System level simulators are becoming more popular. Often these programs are designed to be integrated with schematic capture and PLD development software. These simulators allow not only the simulation of the PLD but also the simulation of the role the PLD will play in the overall circuit.

The practicality of system level simulation depends upon a large number of factors. Theoretically, virtually any system can be simulated with many of the tools available today. In practice, however, the effort to do the simulation may far exceed the effort to build and test the final product!

This question of the relative effort in using simulators brings up some interesting questions. The major purpose of all simulation is, of course, design validation. The simulator allows the design (or at least part of it) to be developed and validated without the need to commit to expensive hardware. For the simulator to make sense economically, it must cost significantly less to run the simulation than it does to build the actual hardware.

The key to cost-effective simulation is the ease of use of the simulator and the availability of device models. Building the necessary models can account for a major cost of the overall simulation. When using dedicated programmable gate array software such as Xilinx's XACT, the device models are included. When trying to perform system level simulation on designs using PLDs however, it is quite likely that any programmable logic will need to be modeled individually.

The functionality of the design is the major concern when doing simulation. However, simulation also allows the evaluation of design under conditions that are difficult or impossible to test otherwise. For example, it may be nearly impossible to physically test a circuit for both the worst case variance in individual tolerances and the maximum temperature extremes. With a simulator, however, a wide variety of such conditions can be expeditiously evaluated.

Mixed signal (often called mixed mode) simulators are becoming more capable and therefore more common. These simulators provide the capability of simulating both analog and digital systems. This type of simulation is quite important in many of today's designs. The engine control system presented in Chapter 11 is an example. It is possible to partition such a system into analog, digital, software, and mechanical elements. Such a partitioning is rather artificial however. For the engine to run smoothly and efficiently, all of the elements must function together smoothly. This cannot and will not happen by trying to develop each element in isolation. The analog, digital, software, and mechanical aspects of the system must all be considered and carefully integrated. Each element is interdependent. Only by viewing the entire system as a whole can such a product be efficiently produced.

These varied disciplines place a heavy burden on any simulator. As simulators continue to improve, their use in the evaluation of such systems will continue to expand. While such simulators are quite expensive and relatively rare today, undoubtedly they will find their way into common use over the next several years.

12.4.3 Device Specific Simulators

Device specific simulators are typically restricted to microcontrollers. These tools simulate, in software, the registers and other resources of the microcontroller. Simulators load the object code and interpret each instruction as if it were actually being executed. Like monitors, simulators allow tight control of, and visibility into, the execution of the program.

In some regards simulators are superior to monitors, yet they are inferior in other aspects. First, the negatives: Since they are only software, simulators cannot interface with real inputs. All inputs or outputs of the simulated microcontroller must also be simulated. Further, simulators do not run in real time and generally will be much slower than the actual microcontroller.

For the positives, simulators can do things that are virtually impossible for monitors or other development tools. For example, a simulator often makes it an easy matter to determine exactly how many machine cycles a particular program sequence takes. This can be invaluable in some real time applications. Simulators can also make program contingency studies much easier. The ability of the simulator to keep track of machine cycles and instructions that have been executed makes it relatively simple to study a wide range of "what if" scenarios.

Another major advantage of simulators is the fact that no hardware is required. This is handy when doing algorithm development or when simply trying to learn the instruction set of a microcontroller. Often, much of the tricky and software intensive code development can be done without ever getting near a real microcontroller!

Finally, we present a general comment on simulators. In Zen, there is a saying that "the map is not the territory." While the meaning of this may be somewhat metaphysical in most aspects, it can be taken quite literally in regard to programmable logic design. The simulation only provides an *approximation* of how the actual programmed circuit will respond. This approximation may be very close or it may differ radically from the actual circuit performance.

Simulation is to be encouraged wherever it is a viable alternative, but it must be remembered that there is no substitute for real world testing.

12.5 In-Circuit Emulators

In the discussion on program monitors, it was noted that these tools have some limitations. First, they use up resources for the development effort. Second, they can only be used with the more sophisticated microcontrollers. In order to overcome these two limitations *in-circuit emulators* were developed. Originally, emulators were limited to microcontrollers and microprocessors. As sequencers and PLDs have become more sophisticated, the availability of emulators for these devices has increased as well.

The basic idea behind the emulator is simple. A general purpose computer such as a PC is connected to a special pod. The pod is configured in such a way that it can be plugged directly into the circuit. The pod is a pin-for-pin and functional equivalent of the device being emulated. A typical arrangement is shown in Figure 12-2.

The emulator acts in much the same manner as the monitor program. However, since the emulator is plugged directly into the system, there is no need to use a serial channel link. The same capability to

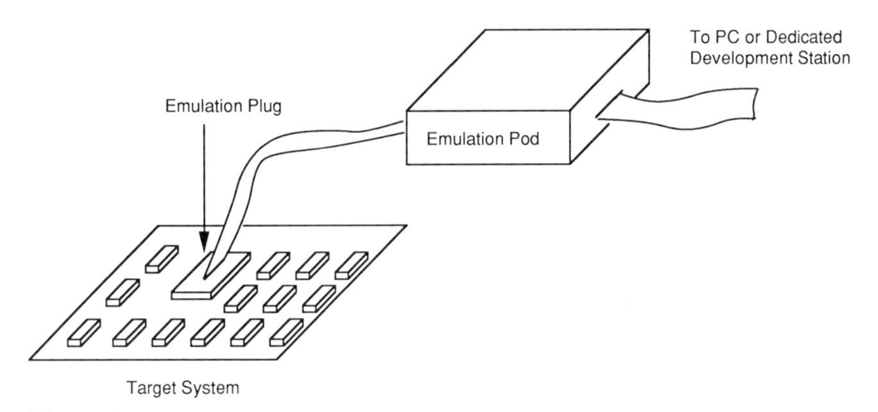

Figure 12-2
Typical Emulator Arrangement.

execute programs, single step, set break points, etc. are available. However, these functions are often implemented with hardware in the emulator. This frees the resources of the microcontroller and makes the code development effort more transparent to the programmer.

Emulators typically emulate all of the functions of the controller, including the program store. Thus, an emulator can be used to develop code for simple microcontrollers that do not have microprocessor-like capabilities. In fact, for many microcontrollers an emulator is the *only* way to develop code.

Emulators are generally expensive tools, though usually quite cost-effective. PC-based emulators can be bought for popular microcontrollers for $1,500 to $5,000. More sophisticated dedicated emulators can cost as much as $50,000.

In comparing monitors to emulators, it may be difficult to determine which development approach is better. The decision depends upon a number of variables. Monitors are far cheaper than emulators and typically provide the same basic capabilities as a true emulator (i.e., single stepping, break points, and on-line code changes). On the downside, monitors are only applicable to relatively sophisticated microcontrollers and require at least some of the microcontroller's resources. Using a monitor requires a greater understanding and more care than does the use of an emulator.

The emulator is the most reliable and least intrusive tool for code development. For simpler units the emulator is the only practical approach to development. Emulators are relatively high-priced items however. This tends to lock one into using a particular type of microcontroller. The cost of having multiple emulator capability for a variety of microcontrollers is often prohibitively expensive.

If the *capital cost* of the development effort is a major concern, the decision is strongly weighted in favor of using monitors as the code development vehicle. Monitors are quite capable of doing everything necessary during the program development effort. On the other hand, monitors are not as efficient or powerful as emulators. For complex programming efforts, the cost of the time lost using a monitor may exceed the money saved by not using an emulator.

On the other hand, a simulator does not really compete as a development alternative. A simulator, no matter how good it is, is not a replacement for either a monitor program or an emulator. The advantages of a simulator, however, are the same regardless of the other development tools used. Thus, a device specific simulator is rarely required but is often a handy tool to have available. If the option of using a simulator is available, it should generally be exercised.

Finally, microcontrollers are continuously growing more sophisticated. For many applications, a monitor program can be included in the architecture with little or no impact on the overall cost of the product. The monitor can greatly help with production testing and field checkout. It therefore becomes an integral part of the product rather than just a development tool. The overall trend is definitely toward such "self-supporting" products.

12.6 Chapter Summary

Development software (i.e., the assemblers, compilers, and other "code generators") are only some of the tools needed to work with programmable logic.

- The basic tools, meters, oscilloscopes, etc. are important to efficient checkout. Their availability and use should not be underrated.
- The logic analyzer is the strongest tool for checking out programmable logic hardware. Logic analyzers monitor the inputs and outputs of the various parts of the system. The monitored data can be formatted and displayed in a variety of ways.
- Monitor programs are simple programs that allow interactive communication with a microcontroller. Monitors allow programs to be downloaded, examined, and executed.
- Simulators allow a design to be evaluated and checked out before committing to hardware. Simulators come in a wide variety of config-

urations. General purpose analog, general purpose digital, and special device simulators are all available.

- Simulation, when practical, is almost always cost-effective, particularly when it is important to evaluate a design over conditions that would be difficult to test in the real world.
- In-circuit emulators allow a general purpose computer to be plugged into a slot that would normally be occupied by an emulated device. This provides visibility into the system during operation and allows the most efficient debugging of system software and hardware.
- In-circuit emulators are available for most microcontrollers and for some higher performance programmable gate arrays.
- In-circuit emulators are mandatory for working with the simpler microcontrollers.

13

A Guide to Choosing Programmable Circuits

Deciding on the "best approach" to designing a circuit is a matter of balance. Some of the variables that need to be considered include:

- Design time.
- Development costs.
- Production cost of the product.
- Learning curves for both design and production teams.
- Available resources.
- Reliability of the final product.
- Maintainability of the final product.
- The ability to extend the product to new applications.

All of these factors will, to a greater or lesser degree, determine the optimum approach. The purpose of this chapter is to provide some general guidelines to show how field programmable devices can help in arriving at viable design solutions.

The first thing to take into consideration is the volume, in units, of the application. This factor will vary depending upon which phase of the product development cycle one is in.

13.1 Proof-of-Concept Phase

In the earliest stages of product development, the design is likely to be for a *proof-of-concept vehicle*. These are simply units that, as the name implies, are built to show that a concept is viable and tenable. Often, the proof-of-concept vehicle is used to help clarify and finalize the initial product specification.

Proof-of-concept vehicles are, by definition, limited to one or two units. Thus the volume is low, material costs are relatively small, and flexibility is paramount. These factors all push the design selection decisions toward maximum flexibility and ease of design.

For simple combinatorial circuits, the PLA or sophisticated PAL devices are necessary. Even if these devices are overpowered, the flexibility and ease of use warrant their selection. The PAL22V10 or the EP300 and other such devices are good candidates. These devices are sufficiently flexible that ongoing changes during the design and checkout phase can generally be accommodated.

Where sequential logic circuits are required, the first question is "how fast must the circuit operate?" For high-speed circuits, PROM-based state machines will generally be the first choice. The PROM-based architecture provides maximum flexibility and is the easiest to understand and modify. Simple combinatorial circuits such as counters can be implemented with registered PLDs, but anything requiring complex waveform generation or sequencing should be in PROM-based architectures. The ease of modification combined with the flexibility of this architecture will more than make up for any extra parts required to implement the design.

Where speed is not critical, microcontrollers or even microprocessors should be chosen. The exact one will depend upon the application, personal preference, and the available development resources. In any case, however, the high-end devices should be selected, preferably ones with EEPROM as the program store.

The use of top-of-the-line units will provide the maximum flexibility and ease of use. This will allow the inevitable changes, design tweaks, and modifications to be implemented in the most expedient fashion. For example, even if the final application is likely to contain a Motorola 6804 or 6805 (which are simple four and eight bit units, respectively), the proof-of-concept vehicle should probably be built around the high-end 68HC811.

To sum things up for the proof-of-concept phase, the objective is to have the maximum flexibility while minimizing the design and debug time. This can generally be achieved by choosing the most sophisticated devices available.

13.2 Preproduction Units

The preproduction units are typically designed to work out any potential production problems *before* committing to full production. Preproduction unit quantities may vary from one or two units to several hundred units.

The priorities of preproduction units are generally material cost and minimizing assembly and test time. In general, the idea is to choose the least expensive approach that will do the job. Factored into this decision must be intangibles such as how much the design is likely to change and how much flexibility is required for future expansion.

It should always be kept in mind that the cheapest place to solve a problem is early in the design of the preproduction units. This creates a somewhat contradictory set of guidelines: Maintain the maximum flexibility so that later changes can be accommodated easily; and restrict the design to only those features that are required to conform to the design requirements.

Finding a balance can be difficult and must be based on factors unique to each design. How stable is the design specification for the proof-of-concept vehicle? Are the customers likely to demand changes early in the product's life cycle? Is the overall production environment stable? Are parts likely to become obsolete and need replacing during the product's life cycle?

The more variability in the above answers, the more the flexibility of the proof-of-concept vehicle should be retained. On the other hand, if the product is well defined and changes are unlikely, now is a good time to aggressively cut the material costs of the product.

The simplest PLDs that can do the design should be chosen. The idea is to find a device that has just enough programmable capability to realize the function. Extra terms that are not utilized are essentially wasted resources. Registered parts should not be chosen if purely combinatorial devices will do.

Study the partitioning of the logic. If mixed combinatorial and registered functions are spread around, try to isolate the combinatorial functions from the registered functions. This allows the lower cost combinatorial PLDs to be used, reserving registered parts for those applications where they are required.

In following these guidelines, it is not enough to simply make decisions on a chip-by-chip basis. One moderately expensive PLD may be cheaper and more reliable than three inexpensive devices. This is particularly true in systems where a large amount of decoding is done. Often, the same common terms are duplicated in several different devices. One large PLD may not need to duplicate these terms and thus will be more efficient.

PROM-based state machines or microsequencer-based control systems cannot be beat for their speed/flexibility product. However, the cost of implementing these architectures is quite high when compared with monolithic solutions. For preproduction and production systems, the monolithic programmable sequencers, such as those discussed in Chapters 6 and 8, should be considered. These devices can dramatically cut the number of components required to implement a design.

Wherever possible, the functionality of the system should be concentrated in a microcontroller. The large program store of these devices, combined with their computational power, generates the maximum capability for the minimum cost. When choosing a microcontroller, remember that there is an inverse relationship between development cost and the sophistication of the microcontroller. The less expensive the microcontroller's units cost, the more development resources and time that are required to use it.

Carefully evaluate what is required in terms of processing speed, interrupt response, memory size, and on-board peripherals. For high-volume, cost-sensitive applications, processing speed or ease of programming are not generally a major concern. Thus simple 4 bit microcontrollers are the preferred option. On the other hand, if large amounts of character manipulations are to be done, an 8 bit solution may be the only practical choice. For mathematically intensive applications, the cost of a 16 bit unit may well be justified.

For all but the most cost-sensitive applications, a good rule of thumb is to strive for approximately 75% utilization of memory. This is true for both program store and data store. This reserve capacity is adequate for accommodating most last minute changes and market driven updates. If more than 75% of the memory is used at the preproduction stage, a larger capacity device is probably a good investment.

The utilized bandwidth of the device should also be evaluated. For simple PLDs, a derating factor of 50% is not unreasonable. For more complicated devices like microcontrollers or microsequencers, it is harder to evaluate. The justification for using these more sophisticated devices is often predicated on the assumption that they will be heavily utilized.

Microsequencers can often be pushed to 90% of their processing capability. Microcontrollers, on the other hand, should observe the 75% utilization rule. For systems making use of multi-tasking, the percentage of bandwidth utilization should be closer to 50%.

13.3 Production Systems

Large production runs have always tended toward either stock devices, such as ROM-based microcontrollers and gate arrays, or custom ASIC devices designed for that particular application. The attraction of these devices are their low per unit cost and for some devices their high performance.

All of these non–field programmable devices represent relatively large capital outlays. Once this outlay has been made, the devices must either be sold as part of a product or scrapped. The chances of successfully adapting such a fixed device to a new application are small.

Even worse are the large lead times associated with these approaches. Masked ROMs or custom ICs cannot be changed in a responsive fashion. Even if the change is small, the processes of manufacturing new devices will typically take months.

For these reasons, even a minor design change can necessitate $5,000 to $50,000 in incurred costs. For very large volume applications, the cost savings associated with these fixed application ICs will continue to predominate. However, this advantage is declining as the cost of PLDs and EPROM-based microcontrollers decreases.

For small and even medium production runs, the cash flow advantages of using field programmable logic often outweigh the cost advantages of ROM-based or ASIC-based solutions. Programmable logic can usually be ordered as needed; it is not necessary to stockpile large inventories. Even if a manufacturer does get stuck with an inventory of programmable devices, the options are much wider. Often the same devices can be used in other products. Or it may be possible to either return or resell the programmable devices. These options suggest that careful attention should be paid before eliminating field programmable devices from a design, even in the production phase.

Generally, much of the margin that was put in at the preproduction phase will be used by the time a product reaches full-scale production. Where this is not the case, it may be reasonable to look at relaxing the requirements of certain parts. As the data base on the product's reliability and performance grows, it may be possible to specify slower parts, narrower temperature ranges, less flexible PLDs, and microcontrollers with smaller memories.

It is in making these kinds of adjustments that the flexibility of field programmable devices truly shine. Different devices from the same family and sometimes different types of devices can often replace each other. This allows the production team the liberty of shopping for the device that will provide the maximum performance at the minimum price.

13.4 Chapter Summary

Deciding on the best approach to developing a circuit is a matter of balance. Relative proficiency with certain design approaches, schedule, cost, and resource availability must all be weighed against time-to-market, reliability, extensibility, and maintainability.

- Unit volume is a key ingredient in choosing an approach. The smaller the number of units being made, the more sophisticated the devices should be. This minimizes the primary cost: design time.
- For large volumes, design time will be amortized over the larger number of units and therefore will be a smaller cost element. In this case, it will generally be worthwhile to design with simpler and less expensive devices, thus minimizing material costs.
- The objective in producing a cost-effective design is to choose the devices that will do the job with the least wasted resources.
- On the other hand, the earlier one is in the design phase, the more margin one should keep!
- One of the best features of field programmable circuits is the flexibility they allow in stocking. Where possible, try to standardize on a small number of devices that can be used in different parts of the design and in different designs. This of course must be balanced against choosing the simplest devices for the job.
- Finally, if "it can be done in software, do it in software!" Software changes are simpler and cheaper, even though the software development effort is likely to be the most expensive part of the design.

14 Conclusion

The overall art of digital design can be logically ordered in hierarchy. One such ordering is shown in Figure 14-1. At the top of the hierarchy is the *system,* typically a general purpose computer.

The general purpose computer is characterized by the fact that it can be programmed in a high-level language; it can run applications programs; and, most importantly, it interfaces directly to the user.

Down one level is the central processing unit (CPU). Two architectures are common for the CPU: the microprogrammable bit-slice circuit and the RISC architecture. Both architectures can be manipulated by the user at the assembly language level, although this is often discouraged for RISC machines.

For the microprogrammable machines, the instruction set itself is programmable and definable. These microprograms are not available to the user but are the domain of the machine's designers. The states, sequences, and control patterns making up the microprogram are all stored in high-speed PROMs and executed by microsequencers.

Next down is the microcontroller level. Often used in stand-alone applications such as handheld instrumentation or engine controls, the microcontroller is also used in disk drives, tape drives, and for off-loading the I/O processing from the CPU. The microcontroller is normally programmed in assembly language, though high-level programming languages are gaining in popularity.

Next in the hierarchy comes the state machine. State machines are found in bus arbiters, waveform generators, counters, etc. State machines may borrow the microprogramming techniques of the microsequencer or they may be implemented by using special PLDs.

An Arcane History of a Few Acronyms

As is the case with many disciplines, programmable logic design has developed its own lexicon of arcane terms. Most of these are borrowed from either electronic engineering or computer science, but a few are unique to programmable logic design.

Before simply defining the terms and acronyms, some background clarification is needed. First, the term *PLD* for programmable logic device is a generic term that covers PALs, PLAs, FPGAs, and other such devices. Technically, it could cover devices such as ROM-based state machines, microsequencers, and microcontrollers. However, by convention, these latter devices are not usually classified as PLDs.

This is also true of the general term *programmable logic*. Usually programmable logic refers to PLDs, which is why this book uses the more inclusive term *programmable circuits* to cover the spectrum of devices.

Another confusing term is *programmer*. When used in reference to a person it means one who writes the programs. The term is also used to describe the hardware that actually loads the pattern or program into the physical IC. As a general rule, programmers are involved with applications code or systems programming on general purpose computers. Engineers are normally chosen for programming microcontrollers and other programmable circuits. As programmers (the human type) become more hardware-oriented and engineers become more software-oriented, this distinction is blurring.

Even more obscure is the rationale behind terms like *ROM* and *RAM*. Historically it makes some sense, but without the history there would appear to be no rationale at all in these naming conventions. Here is a condensed history.

Initially, memory was composed of two basic types: sequential access memory [tapes (paper or magnetic), drums, and disks] and random access memories (mainly magnetic cores). The terms make sense since they describe the way the memory is accessed. Sequential memory must read all of the data in sequence to get to the data that is desired. A magnetic tape, for example, must start at the beginning of the tape and read until it gets to the data it needs. Magnetic cores, on the other hand, are formed into three-dimensional matrixes. The data can be obtained by simply providing the coordinates of the data. Only the information needed was read—there was no need to read through any other data. Thus the term *random access memory*.

So much for the background. When semiconductor memory came along, it mainly replaced RAM circuitry. Thus RAM became synonymous with the current usage: volatile, semiconductor memory that can be read and written repeatedly.

In some applications there was a need to store programs permanently. In the older computers this was done on tape. The data was read from the tape into RAM, where the program was actually executed. The relatively low cost of semiconductors made it possible to skip this loading sequence. The program could be "wired" directly into the memory chips when they were manufactured. This eliminated the need for the tape drive altogether in many applications. Of course, since the data was wired into the chip, it could only be read. It was not possible to alter the data by writing to it. Thus, the term *read only memory* (ROM) was coined to differentiate this type of semiconductor memory from the more conventional read/write memory (i.e., RAM).

However, a ROM is still technically a RAM. The data can be accessed randomly. It is just convention that differentiates the two.

The use of ROM was handy but not *that* handy. Any change to the program required a complete new manufacturing run of ICs. Not only was this expensive but changes were glacially slow. To overcome this turnaround time and at least some of the cost, the *programmable read only memory* (PROM) was developed. These devices make use of fuses rather than wiring to store data. The fuses could even be blown in the field, making changes relatively easy and cost-effective.

In IC terms, fuses are relatively large. This limits the practical size of fuse-based parts. Worse, once they are blown that is it. The data cannot be changed. Get it right the first time or trash the part.

These factors led to the development of the *erasable programmable read only memory* (EPROM). EPROM technology is discussed in Chapter 3. The main point is that these devices are used in the same way as

PROMs but they can be erased by ultraviolet light. EPROMs can be both read and written, but the read operation is awkward so they are used mostly to replace PROMs. So even though they are read/write semiconductor memory, nobody calls them RAM.

EPROM technology was so successful that many people simply used EPROMs in the final product. This meant that the EPROM was burned once and never erased. Since providing the ultraviolet window in an EPROM is an expensive operation, manufacturers decided to provide EPROMs without the window. This brought things back to a device that could only be written once but never erased. But they are not fuse-based so nobody wanted to call them PROMs. Instead, the nomenclature "one time programmable EPROM" was adopted. Sometimes called ZAT, for zero turn around (I know, but ZTA has no appeal to it, so it was ZAT).

Actually, this is not as crazy as it sounds. Not only are one time programmable (OTP) type parts cheaper but, unlike fuse-based parts, they can undergo 100% testing at the factory.

Erasing an EPROM with ultraviolet light is awkward. It is hard to get the light into a practical application. So there was obviously a need for a nonvolatile memory that could be written to and erased electronically. The result was the electrically erasable programmable read only memory (E^2PROM or EEPROM). This memory can be read and written in-circuit and is of course a semiconductor memory. But it does have a limited number of write cycles before it wears out and is not volatile, so it would be confusing to call it RAM.

Thus we can see that virtually all semiconductor memory is, in fact, actually RAM. But each of the various flavors are differentiated by some other terms. All of this, believe it or not, is to *prevent* confusion.

What this has to do with programmable circuits is twofold. First, microcontrollers and even microsequencers make use of all of these memory technologies. And second, the basic mechanisms have been adopted into the design of PLDs. Often these get renamed in the process. PEEL, for example, is a term used by International CMOS Technology to describe their programmable electrically erasable logic.

The acronyms start flying fast and heavy when the marketing people are striving for product differentiation. And often there is some process difference that the manufacturer points to in order to justify the nomenclature change. But, in general, all PLDs will fall into one of the headings described above. For an exception, see the definition of antifuse in the glossary.

B

Data Sheets

All data sheets presented in this appendix are from the Texas Instruments publication *Programmable Logic Data Book,* 1990 edition. Data sheets are reprinted with permission of the copyright holder, Texas Instruments, © 1990.

TIBPLS506C
13 × 97 × 8 FIELD-PROGRAMMABLE LOGIC SEQUENCER

D3090, DECEMBER 1987 – REVISED NOVEMBER 1989

- 50-MHz Max Clock Rate
- 2 Transition Complement Array Terms
- 16-Bit Internal State Registers
- 8-Bit Output Registers
- Outputs Programmable for Registered or Combinational Operation
- Ideal for Waveform Generation and High-Performance State Machine Applications
- Programmable Output Enable
- Programmable Clock Polarity

JT OR NT PACKAGE
(TOP VIEW)

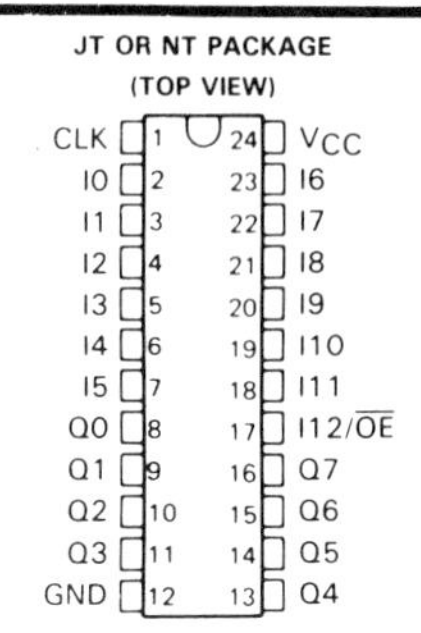

FK OR FN PACKAGE
(TOP VIEW)

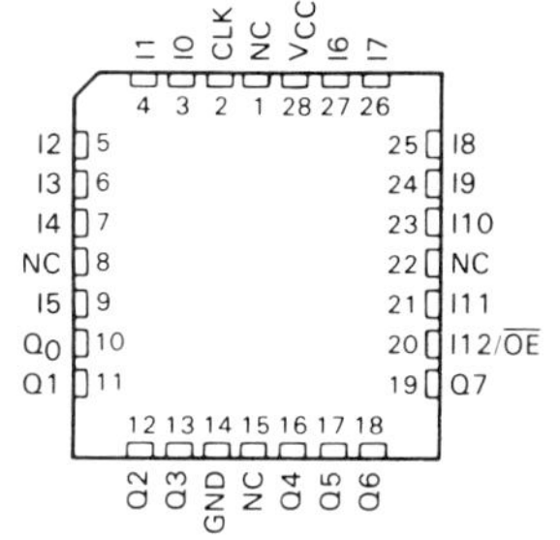

NC – No internal connection

description

The TIBPLS506 is a TTL field-programmable state machine of the Mealy type. This state machine (logic sequencer) contains 97 product terms (AND terms) and 48 sum terms (OR terms). The product and sum terms are used to control the 16-bit internal state registers and the 8-bit output registers.

The outputs of the internal state registers (P0-P15) are fed back and combined with the 13 inputs (I0-I12) to form the AND array. In addition, two sum terms are complemented and fed back to the AND array, which allows any product terms to be summed, complemented, and used as inputs to the AND array.

The eight output cells can be individually programmed for registered or combinational operation. Nonregistered operation is selected by blowing the output multiplexer fuse. Registered output operation is selected by leaving the output multiplexer fuse intact.

Pin 17 can be programmed to function as an input and/or an output enable. Blowing the output enable fuse lets pin 17 function as an output enable but does not disconnect pin 17 from the input array. When the output enable fuse is intact, pin 17 functions only as an input with the outputs being permanently enabled.

The state and output registers are synchronously clocked by the fuse programmable clock input. The clock polarity fuse selects either postive- or negative-edge triggering. Negative-edge triggering is selected by blowing the clock polarity fuse. Leaving this fuse intact selects positive-edge triggering. After power-up, the device must be initialized to the desired state. When the output multiplexer fuse is left intact, registered operation is selected.

The TIBPLS506C is characterized for operation from 0°C to 75°C.

TIBPLS506C
13 × 97 × 8 FIELD-PROGRAMMABLE LOGIC SEQUENCER

logic diagram (positive logic)

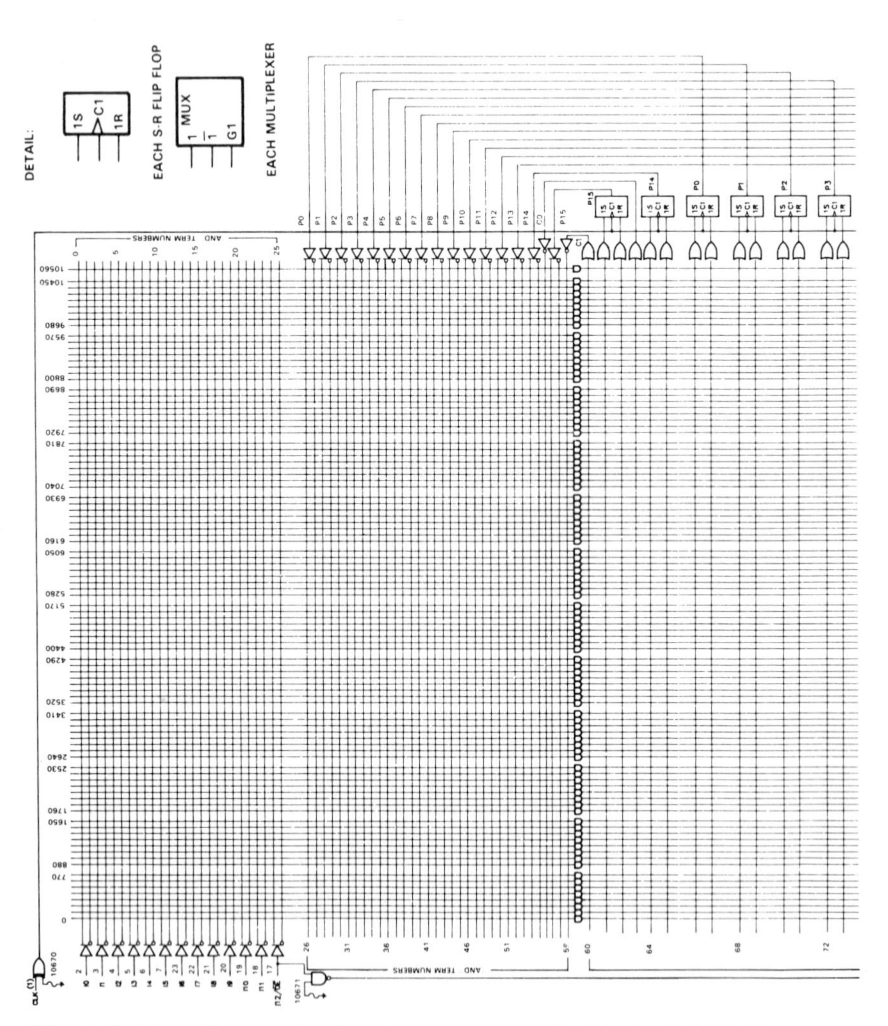

NOTES: A. All inputs to AND gates, exclusive-OR gates, and multiplexters with a blown link assume the logic-1 state.
B. All OR gate inputs with a blown link assume the logic-0 state.

TIBPLS506C
13 × 97 × 8 FIELD-PROGRAMMABLE LOGIC SEQUENCER

logic diagram (continued)

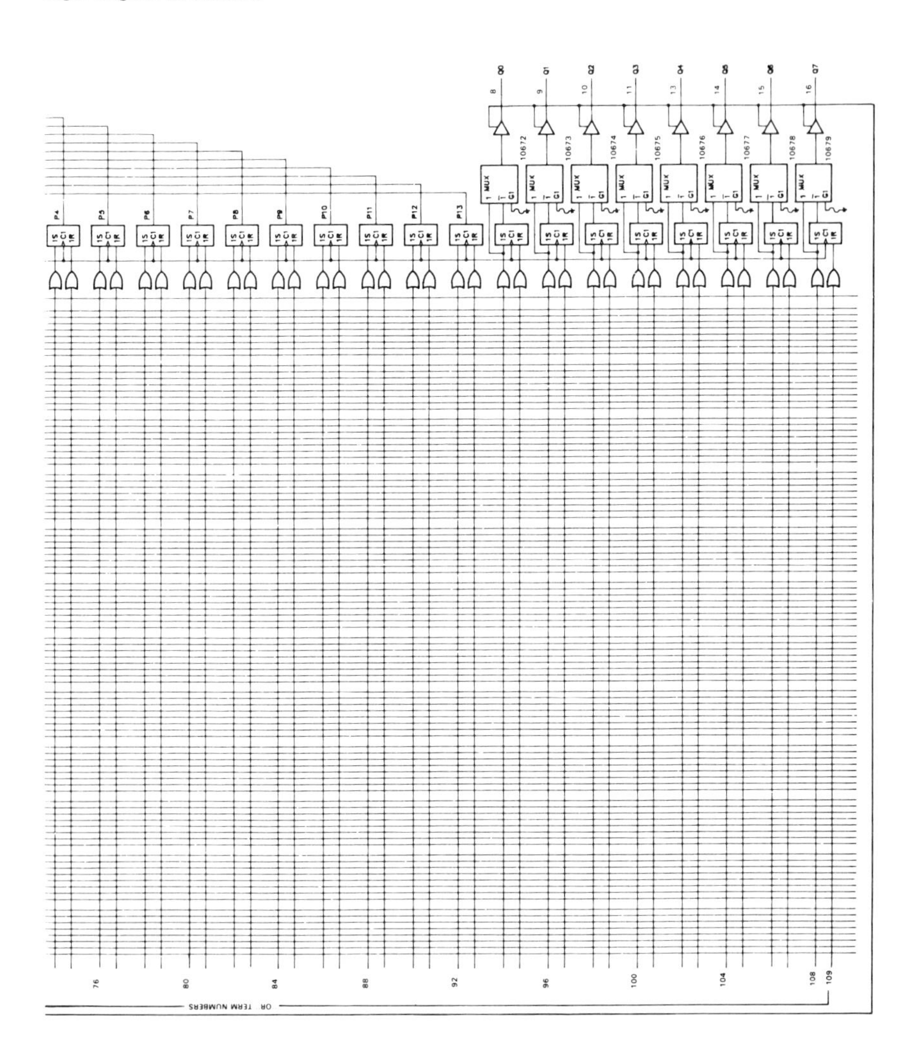

TIBPLS506C
13 × 97 × 8 FIELD-PROGRAMMABLE LOGIC SEQUENCER

S-R FUNCTION TABLE (see Note 1)

CLK POLARITY FUSE	CLK	S	R	STATE REGISTER
INTACT	↑	L	L	Q_0
INTACT	↑	L	H	L
INTACT	↑	H	L	H
INTACT	↑	H	H	INDETERMINATE
BLOWN	↓	L	L	Q_0
BLOWN	↓	L	H	L
BLOWN	↓	H	L	H
BLOWN	↓	H	H	INDETERMINATE

NOTE 1: Q_0 is the state of the S-R registers before the active clock edge.

functional block diagram (positive logic)

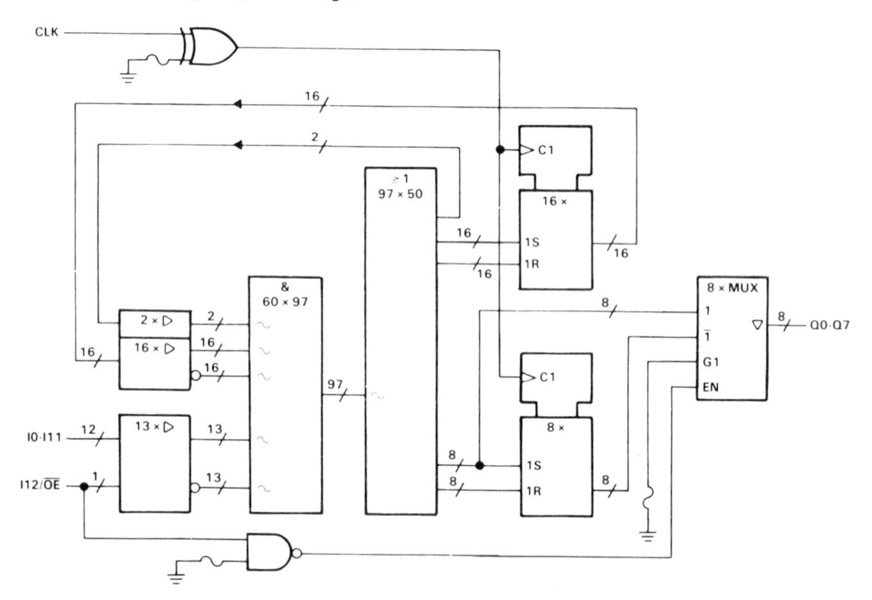

~ denotes fused inputs.

TIBPLS506C
13 × 97 × 8 FIELD-PROGRAMMABLE LOGIC SEQUENCER

absolute maximum ratings over operating free-air temperature range (unless otherwise noted)

Supply voltage, V_{CC} (see Note 2) .... 7 V
Input voltage (see Note 2) .... 5.5 V
Voltage applied to disabled output (see Note 2) .... 5.5 V
Operating free-air temperature range .... 0 °C to 75 °C
Storage temperature range .... −65 °C to 150 °C

NOTE 2: These ratings apply except when programming pins during a programming cycle or during diagnostic testing.

recommended operating conditions

			MIN	NOM	MAX	UNIT
V_{CC}	Supply voltage		4.75	5	5.25	V
V_{IH}	High-level input voltage, V_{CC} = 5.25 V		2		5.5	V
V_{IL}	Low-level input voltage, V_{CC} = 4.75 V				0.8	V
I_{OH}	High-level output current				−3.2	mA
I_{OL}	Low-level output current				16	mA
t_w	Pulse duration	Clock high	6			ns
		Clock low	6			
t_{su}	Setup time before CLK† input or feedback to S-R inputs	Without C-array	15			ns
		With C-array	25			
t_h	Hold time after CLK	Input or feedback at S-R inputs	0			ns
T_A	Operating free-air temperature		0		75	°C

† The active edge of CLK is determined by the programmed state of CLK polarity fuse.

electrical characteristics over recommended operating free-air temperature range (unless otherwise noted)

PARAMETER	TEST CONDITIONS	MIN	TYP‡	MAX	UNIT
V_{IK}	V_{CC} = 4.75 V, I_I = −18 mA			−1.2	V
V_{OH}	V_{CC} = 4.75 V, I_{OH} = −3.2 mA	2.4	3		V
V_{OL}	V_{CC} = 4.75 V, I_{OL} = 16 mA		0.37	0.5	V
I_I	V_{CC} = 5.25 V, V_I = 5.5 V			0.1	mA
I_{IH}	V_{CC} = 5.25 V, V_I = 2.7 V			20	μA
I_{IL}	V_{CC} = 5.25 V, V_I = 0.4 V			−0.25	mA
I_O§	V_{CC} = 5.25 V, V_O = 0.5 V	−30		−130	mA
I_{OZH}	V_{CC} = 5.25 V, V_O = 2.7 V			20	μA
I_{OZL}	V_{CC} = 5.25 V, V_O = 0.4 V			−20	μA
I_{CC}	V_{CC} = 5.25 V, See Note 3, Outputs open		156	210	mA
C_i	f = 1 MHz, V_I = 2 V		7		pF
C_o	f = 1 MHz, V_O = 2 V		11		pF
C_{clk}	f = 1 MHz, V_I = 2 V		14		pF

‡ All typical values are at V_{CC} = 5 V, T_A = 25 °C.
§ This parameter approximates I_{OS}. The condition V_O = 0.5 V takes tester noise into account. Not more than one output should be shorted at a time and duration of the short circuit should not exceed one second.
NOTE 3: When the clock is programmed for negative-edge, then V_I = 4.75 V. When the clock is programmed for positive-edge, then V_I = 0.

TIBPLS506C
13 × 97 × 8 FIELD-PROGRAMMABLE LOGIC SEQUENCER

switching characteristics over recommended ranges of supply voltage and operating free-air temperature (unless otherwise noted)

PARAMETER	FROM	TO	TEST CONDITIONS	MIN	TYP†	MAX	UNIT
f_{max}‡		Without C-array	R_1 = 300 Ω, R_2 = 390 Ω, See Figure 3	50	65		MHz
		With C-array		33	50		
t_{pd}§	CLK↑	Q (nonregistered)		8		27	ns
	CLK↓			9		28	
t_{pd}§	CLK↑	Q (registered)		3		10	ns
	CLK↓			4		11	
t_{pd}	I or Feedback	Q (nonregistered)		10		22	ns
t_{en}	OE↓	Q		2	6	10	ns
t_{dis}	OE↑	Q	C_L = 5 pF	2	6	10	ns

†All typical values are at V_{CC} = 5 V, T_A = 25°C.

‡f_{max}, with external feedback, can be calculated as $\frac{1}{t_{su} + t_{pd}\ \text{CLK to Q}}$. f_{max} is independent of the internal programmed configuration and the number of product terms used.

§The active edge of CLK is determined by the programmed state of the CLK polarity fuse.

timing model

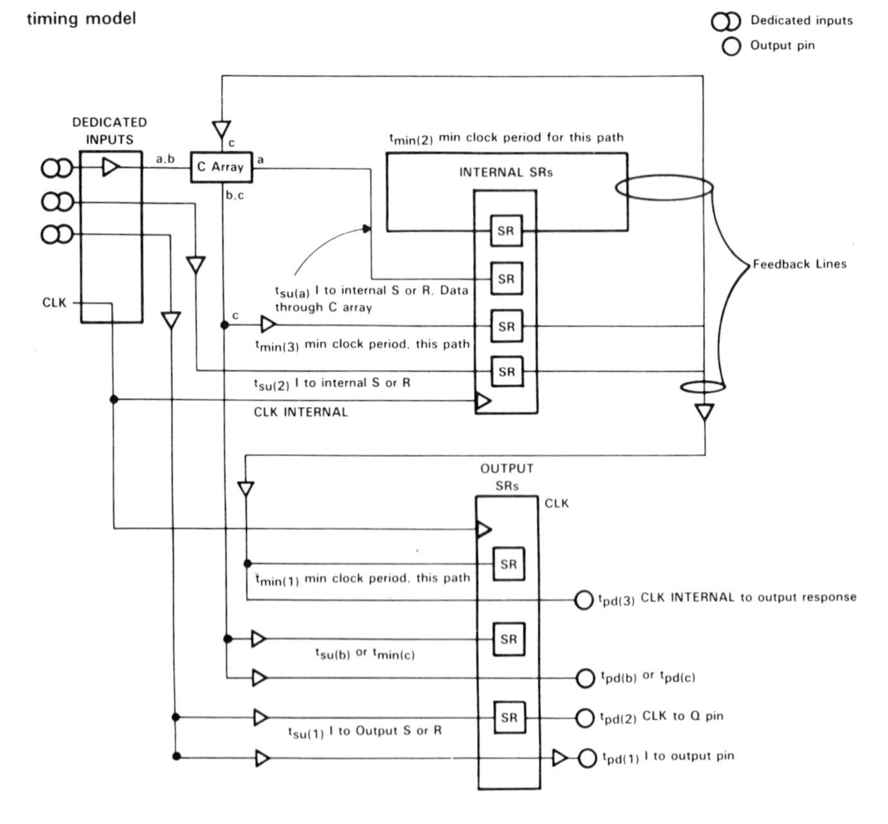

TIBPLS506C
13 × 97 × 8 FIELD-PROGRAMMABLE LOGIC SEQUENCER

glossary — timing model

$t_{pd(1)}$ — Maximum time interval from the time a signal edge is received at any input pin to the time any logically affected combinational output pin delivers a response.

$t_{pd(2)}$* — Maximum time interval from a positive edge on the clock input pin to data delivery on the output pin corresponding to any output SR register.

$t_{pd(3)}$* — Maximum time interval from the positive edge on the clock input pin to the response on any logically affected combinationally configured output (at the pin), where data origin is any internal SR register.

$t_{pd(b)}$ — Maximum time interval from the time a signal edge is received at any input pin to the time any logically affected combinational output pin delivers a response, where data passes through a **C ARRAY** once before reaching the affected output.

$t_{pd(c)}$* — Maximum time interval from the positive edge on the clock input pin to the response on any logically affected combinationally configured output (at the pin), where data origin is any internal SR register and data passes once through a **C ARRAY** before reaching an affected output.

$t_{su(1)}$ — Minimum time interval that must be allowed between the data edge on any dedicated input and the **active** clock edge on the clock input pin when data affects the S or R line of any output SR register.

$t_{su(2)}$ — Minimum time interval that must be allowed between the data edge on any dedicated input and the **active** clock edge on the clock input pin when data affects the S or R line of any internal SR register.

$t_{su(a)}$ — Minimum time interval that must be allowed between the data edge on any dedicated input and the **active** clock edge on the clock input pin when data passes once through a C ARRAY before reaching an affected S or R line on any internal SR register.

$t_{su(b)}$ — Minimum time interval that must be allowed between the data edge on any dedicated input and the **active** clock edge on the clock input pin when data passes once through a C ARRAY before reaching an affected S or R line on any output SR register.

$t_{min(1)}$ — Minimum clock period (or 1/[maximum frequency]) that the device will accomodate when using feedback from any internal SR register or counter bit to feed the S or R line of any output SR register.

$t_{min(2)}$ — Minimum clock period (or 1/[maximum frequency]) that the device will accomodate when using feedback from any internal SR register to feed the S or R line of any internal SR register.

$t_{min(3)}$ — Minimum clock period (or 1/[maximum frequency]) that the device will accomodate when using feedback from any internal SR register to feed the S or R line of any internal SR register and data passes once through a C ARRAY before reaching an affected S or R line on any internal SR register.

$t_{min(c)}$ — Minimum clock period (or 1/[maximum frequency]) that the device will accomodate when using feedback from any internal SR register to feed the S or R line of any output SR register and data passes once through a C ARRAY before reaching an affected S or R line on any output SR register.

TIBPLS506C
13 × 97 × 8 FIELD-PROGRAMMABLE LOGIC SEQUENCER

PARAMETER VALUES FOR TIMING MODEL

$t_{pd(1)}$ = 22 ns	$t_{su(1)}$ = 15 ns	$t_{min(1)}$ = 20 ns
$t_{pd(2)}$* = 10 ns	$t_{su(2)}$ = 15 ns	$t_{min(2)}$ = 20 ns
$t_{pd(3)}$* = 27 ns	$t_{su(a)}$ = 25 ns	$t_{min(3)}$ = 25 ns
	$t_{su(b)}$ = 25 ns	$t_{min(c)}$ = 25 ns

INTERNAL NODE NUMBERS

Q0-Q7	RESET 25-32	P0-P15	SET 33-48
C0	65		RESET 49-64
C1	66		

diagnostics

A diagnostic mode is provided with these devices that allows the user to inspect the contents of the state registers. The step-by-step procedures required to use the diagnostics follow.

1. Disable all outputs by taking pin 17 ($\overline{OE}$) high (see Note 4).
2. Take pin 8 (Q0) double high to enable the diagnostics test sequence.
3. Apply appropriate levels of voltage to pins 11 (Q3), 13 (Q4), and 14 (Q5) to select the desired state register (see Table 1).

The voltage level monitored on pin 9 will indicate the state of the selected state register.

NOTE 4: If pin 17 is being used as an input to the array, then pin 7 (I5) must be taken double high before pin 17 is taken high.

diagnostics waveforms

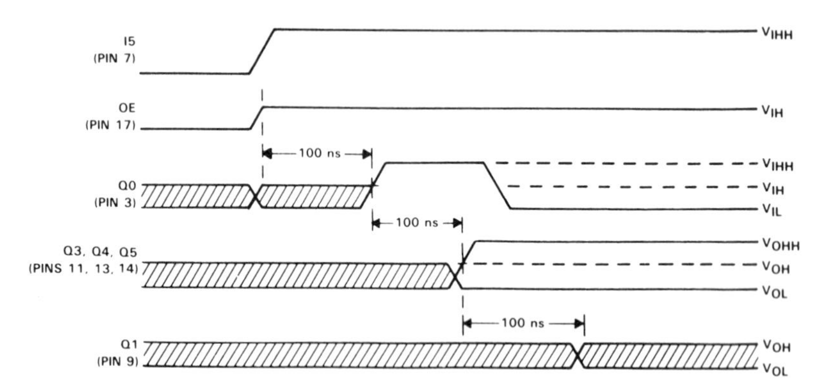

TIBPLS506C
13 × 97 × 8 FIELD-PROGRAMMABLE LOGIC SEQUENCER

TABLE 1. ADDRESSING STATE REGISTERS DURING DIAGNOSTICS†

REGISTER BINARY ADDRESS			BURIED REGISTER
PIN 11	PIN 13	PIN 14	SELECTED
L	L	L	C1
L	L	H	P15
L	L	HH	CO
L	H	L	P14
L	H	H	PO
L	H	HH	P1
L	HH	L	P2
L	HH	H	P3
L	HH	HH	P4
H	L	L	P5
H	L	H	P6
H	L	HH	P7
H	H	L	P8
H	H	H	P9
H	H	HH	P10
H	HH	L	P11
H	HH	H	P12
H	HH	HH	P13

† V_{IHH} = 10.25 V min, 10.5 V nom, 10.75 V max

programming information

Texas Instruments programmable logic devices can be programmed using widely available software and reasonably priced device programmers.

Complete programming specifications, algorithms, and the latest information on firmware, software, and hardware updates are available upon request. Information on programmers that are capable of programming Texas Instruments programmable logic is also available, upon request, from the nearest TI sales office, local authorized Texas Instruments distirbutor, or by calling Texas Instruments at (214) 997-5666.

TIBPLS506C
13 × 97 × 8 FIELD-PROGRAMMABLE LOGIC SEQUENCER

TYPICAL APPLICATIONS

f_{max}

When the TIBPLS506 is used with two or more devices linked to build a ''multi-device'' state machine (see Figure 1), the maximum operating frequency for this state machine is limited to the sum of t_{pd} CLK-Q (10 ns) of the first '506 and t_{su} (15 ns), of the second '506, for a clock period of 25 ns. This results in an f_{max} of 40 MHz (1/25 ns).

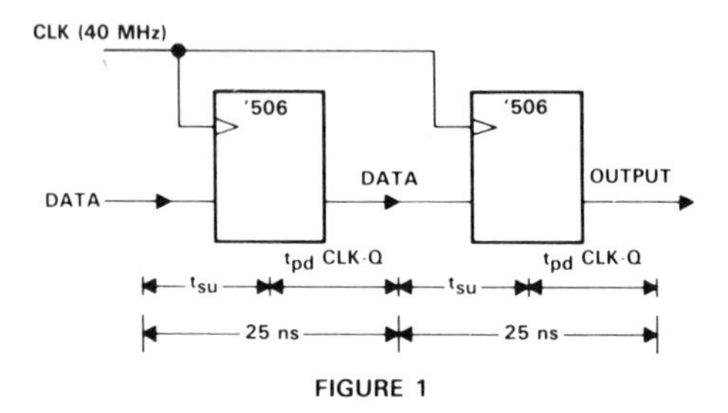

FIGURE 1

Figure 2 shows the '506 used in a system environment where it is operated at 50 MHz, the highest clock rate possible without compromising data integrity. At the input of the '506, the system clock period is limited to the sum of t_{pd} CLK-Q of device A and t_{su} of the '506. At the output of the '506, the system clock period is limited to the sum of t_{pd} CLK-Q of the '506 and t_{su} of device B.

For this system to operate at 50 MHz, a system clock period of 20 ns must be met. Given that t_{su} for the '506 is 15 ns minimum, t_{pd} CLK-Q of device A cannot exceed 5 ns (15 ns + 5 ns = 20 ns). On the output side of the '506, t_{pd} CLK-Q of 10 ns must be allowed. In order to meet the system clock period of 20 ns, t_{su} for device B must not exceed 10 ns (10 ns + 10 ns) = 20 ns). Under these circumstances, a system frequency of 50 MHz (1/20 ns) can be realized.

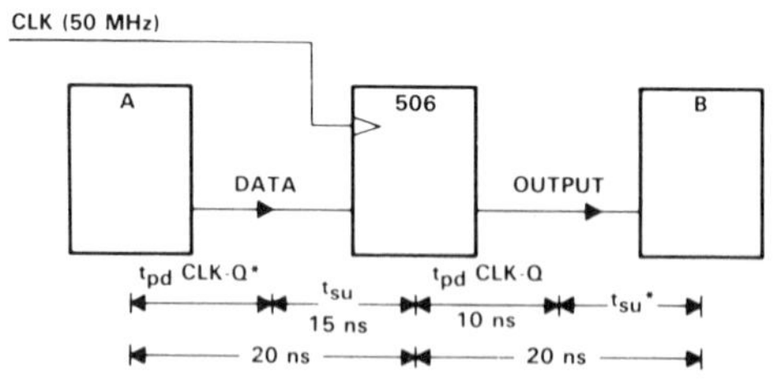

*External device parameters (t_{pd} CLK-Q of device A ≤ 5 ns, and t_{su} of device B ≤ 10 ns)

FIGURE 2

TIBPLS506C
13 × 97 × 8 FIELD-PROGRAMMABLE LOGIC SEQUENCER

PARAMETER MEASUREMENT INFORMATION

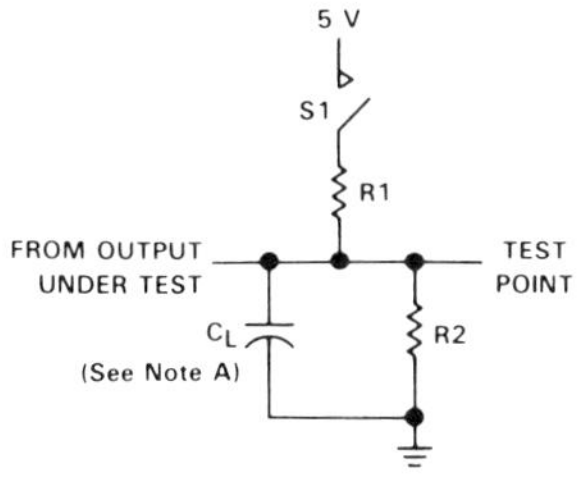

LOAD CIRCUIT FOR THREE-STATE OUTPUTS

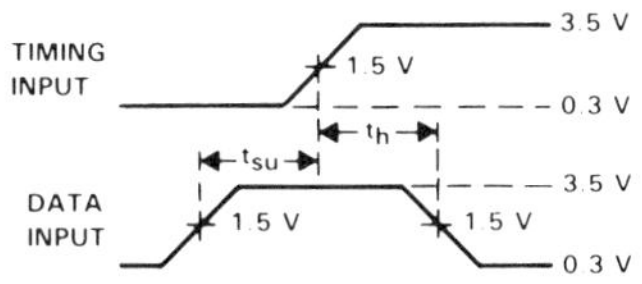

VOLTAGE WAVEFORMS SETUP AND HOLD TIMES

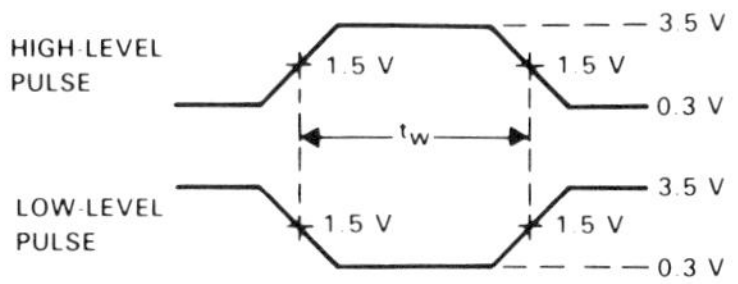

VOLTAGE WAVEFORMS PULSE DURATIONS

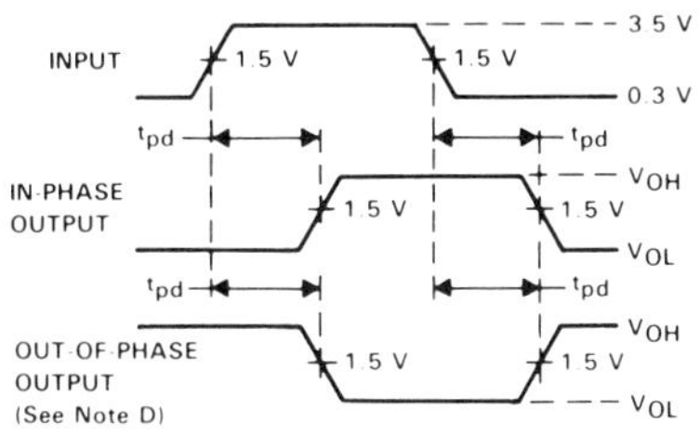

VOLTAGE WAVEFORMS PROPAGATION DELAY TIMES

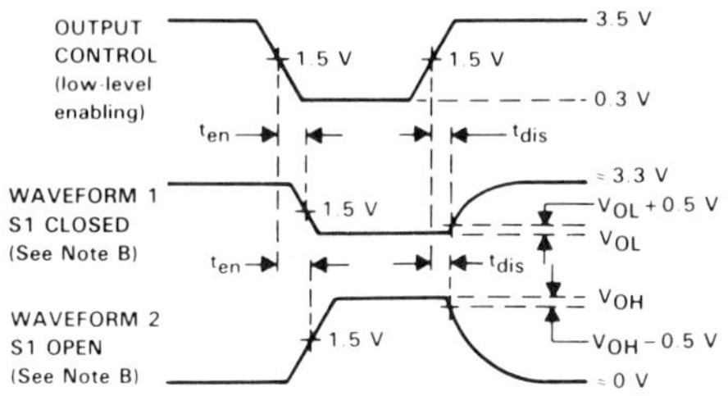

VOLTAGE WAVEFORMS ENABLE AND DISABLE TIMES, THREE-STATE OUTPUTS

NOTES:
A. C_L includes probe and jig capacitance and is 50 pF for t_{pd} and t_{en}, 5 pF for t_{dis}.
B. Waveform 1 is for an output with internal conditions such that the output is low except when disabled by the output control. Waveform 2 is for an output with internal conditions such that the output is high except when disabled by the output control.
C. All input pulses have the following characteristics: PRR ≤ 1 MHz, $t_r = t_f = 2$ ns, duty cycle = 50%.
D. When measuring propagation delay times of 3-state outputs, switch S1 is closed.

FIGURE 3

TIBPLS506C
13 × 97 × 8 FIELD-PROGRAMMABLE LOGIC SEQUENCER

TYPICAL CHARACTERISTICS

SUPPLY CURRENT
vs
FREE-AIR TEMPERATURE

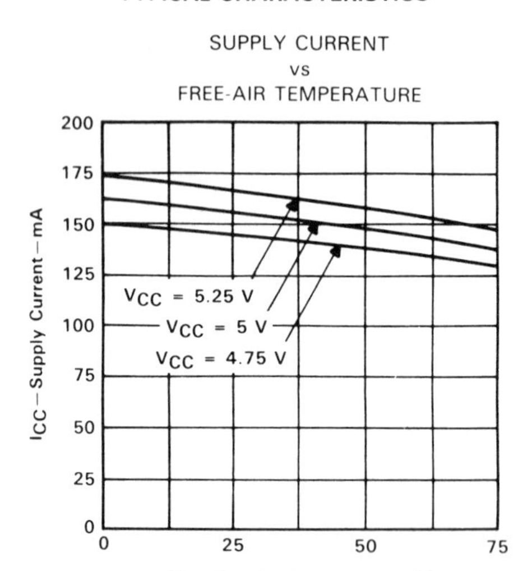

FIGURE 4

POWER DISSIPATION
vs
FREQUENCY

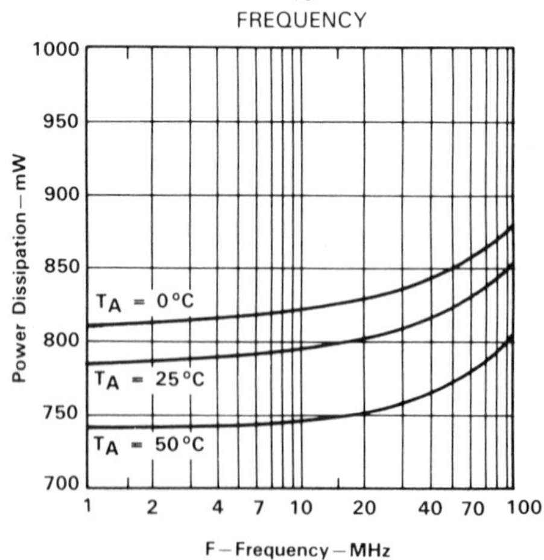

FIGURE 5

TIBPLS506C
13 × 97 × 8 FIELD-PROGRAMMABLE LOGIC SEQUENCER

TYPICAL CHARACTERISTICS

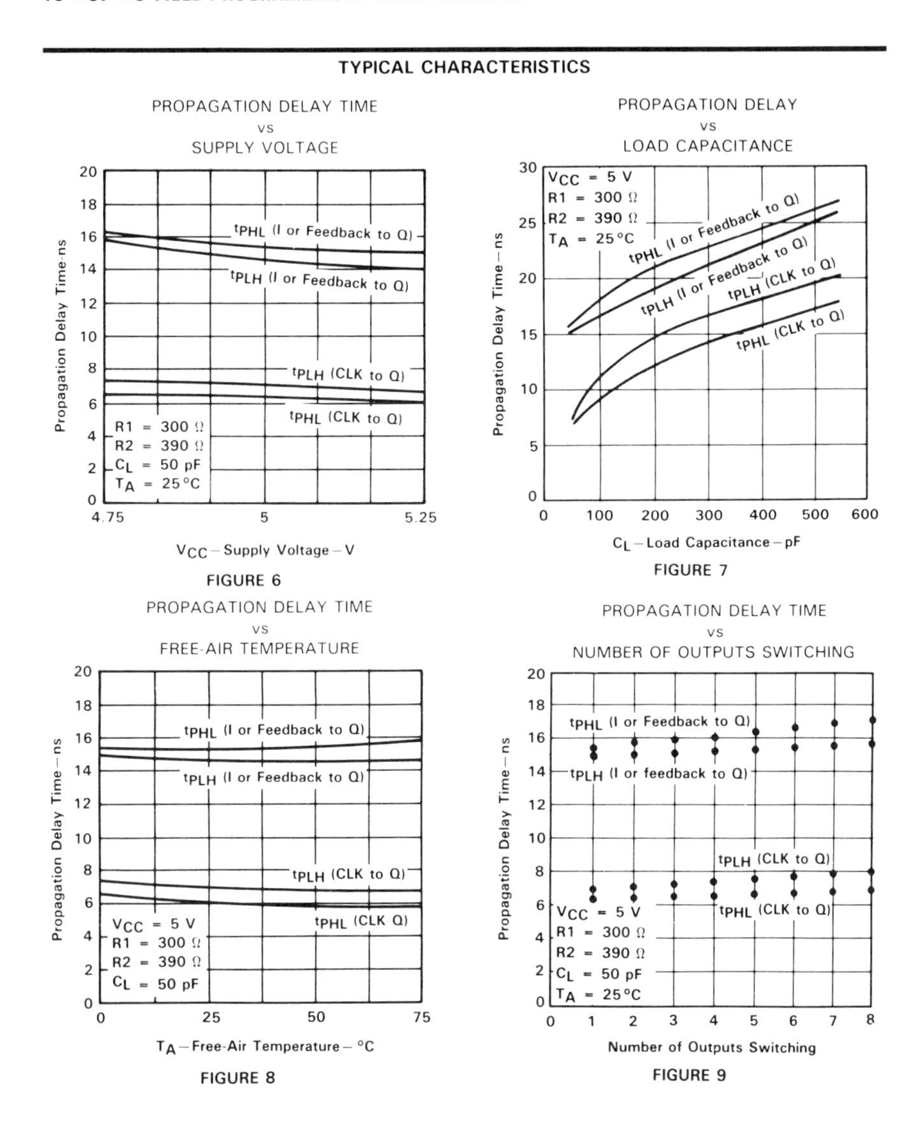

FIGURE 6

FIGURE 7

FIGURE 8

FIGURE 9

C

References and Sources

One of the major problems in keeping up with the rapid growth of programmable circuit design techniques is simply tracking the volume of new material.

This book has focused on laying a groundwork of fundamental principles which will change only slowly with time. For the latest technical information on specific devices, the best sources are the field application engineers (FAEs) provided by most vendors. A few phone calls will often yield many volumes on the latest devices, which is why this book does not devote much space to specific products.

The local FAE can be found by calling the regional office of the vendor of interest or by asking the local electronic distributors for contacts.

Specific books and articles that have been of use in this work include:

Advanced Micro Devices, *PAL Device Handbook*. Advanced Micro Devices, Sunnyvale, California, 1988.

Alford, Roger C., *Programmable Logic Designer's Guide*. Howard W. Sams & Company, Carmel, Indiana, 1989.

Foster, Caxton C., *Real Time Programming—Neglected Topics*. Addison-Wesley, Reading, Massachusetts, 1981.

Leibson, Steven H., PLD development software. *EDN* (August 2, 1990), pp. 100–116.

Marino, Leonard R., *Principles of Computer Design*. Computer Science Press, Rockville, Maryland, 1986.

Mellichamp, Duncan A. (Ed)., *Real Time Computing*. Van Nostrand Reinhold, New York, 1983.

Motorola Inc., *M68HC11 Reference Manual*. Prentice-Hall, Englewood Cliffs, New Jersey, 1989.
Motorola Inc., *M68HC05 Microcontroller Applications Guide*. Motorola, Inc., 1989.
National Semiconductor, *Programmable Logic Design Guide*. Sunnyvale, California, 1986.
Quinell, Richard A., JTAG boundary scan test: Adding testability also aids debugging. *EDN* (August 2, 1990), pp. 67–74.
Robie, Jonathan, Fair share. *Byte* (July, 1988), pp. 229–236.

Glossary

Accumulator A special register or group of registers that are used for data manipulation and arithmetic operations. Normally found in microcontrollers, microprocessors, and general purpose computers.

Active High A logic signal that when near the positive supply voltage (typically 5 V) is active. Most logic signals are defined as active high simply because it is easier to conceptualize active-high signals.

Active Low A logic signal that when near ground potential is in the active state. Active-low signals are normally used to enable ICs.

AND The logical operator that is true if and only if all of its inputs are true.

Antifuse A technique for making programmable connections in a logic device. The antifuse is essentially a capacitor that is programmed by applying a voltage above the breakdown voltage. The capacitor will short, thus making a connection between two points in the circuit. This is just the opposite of a fuse and thus the name.

Assembler A program that is used to convert symbolic information to fuse patterns (PLDs), microcode (meta-assemblers), or object code (microcontrollers).

Assembly Language The mnemonics, operations, and operators for the machine level instructions used by a microcontroller.

Bandwidth Technically, this is an analog rather than digital term. However, the term is often applied to mean the amount of information that can flow over a data bus or the amount of data that a processor can handle in a given period of time.

Bit Contraction of binary digit. A bit is either a zero or a one. Early electromechanical computers represented a "zero" by an open relay and a "one" by a closed relay. Thus, a "one" is often referred to as "CLOSED" or "ON" and a zero as "OPEN" or "OFF." These terms are discouraged since they are misleading but the practice is a common one.

BIT Acronym for built-in test. Refers to either hardware or software features or products that are built in to allow the detection and isolation of failures.

Bit-Slice A term applied to building a central processing unit from slices. For example, four 4 bit slices would be used to build a 16 bit accumulator. Bit-slice circuits are usually controlled by special state machines known as microsequencers.

Boundary Scan A technique for improving system testability by adding registers around the boundary of a chip, board, or system. The techniques have been formalized by the Joint Test Analysis Group (JTAG).

Branch Instruction An instruction that causes the program to transfer control to other than the next instruction in line.

C A high-level language (HLL) that is often used for system programming and other hardware intensive tasks. C is the most popular HLL for microcontrollers, though most are programmed in assembly language.

CAD Acronym for computer-aided design.

CAM Acronym for computer-aided manufacturing.

Canonical Form An equation is said to be in its canonical form if it is expressed in either the standard sum of products (SOP) or the standard product of sums (POS) form.

CASE Acronym for computer-aided systems engineering.

Chip Slang term for an integrated circuit.

CISC Acronym for complex instruction set computer. Most modern computers such as PCs and VAXes are CISC architecture machines.

Compiler A program for translating symbolic information to a bit pattern (PLDs) or object code (microcontrollers). Compilers allow a greater degree of abstraction and more flexibility than is allowed by an assembler. Object code produced by a compiler is often not as efficient as that produced by using an assembler.

Control Store The portion of memory that is used to store either the program (microcontrollers) or the microprogram (microsequencers).

CPU Acronym for central processing unit. The CPU contains the accumulators, memory control circuitry, and other specialized functions necessary for logical and arithmetic operations. All data normally passes through the CPU.

CRC Acronym for circular redundancy check. A formalized method of using a polynomial to continuously divide a bit stream. The remainder is then appended to the bit stream. The point in all of this is that if the bit stream is sent through the CRC process again, the remainder should equal zero. If it does not, the bit stream has been corrupted.

Cross Assembler, Cross Compiler The term *cross* is used to designate that the assembler or compiler is generating code for a machine other than the machine that is doing the assembly or compilation. See *native code*.

Data Store Typical RAM, the data store is used to store variables, working values, etc.

De Morgan's Theorem The product of two variables is equal to the complement of the sum of the complemented variables.

Development Station A general purpose computer used to edit and assemble source files, download programs, etc.

Downloading The process of moving the object code from the development station to the target system.

EEPROM Acronym for electrically erasable programmable read only memory. EEPROM memory can be written to, but the process is much slower than reading. However, once written the data is retained even when power is removed.

EPROM Acronym for electrically programmable read only memory. EPROM memory can be written to, but usually only by a hardware programmer. Before new data can be written, the old data must be erased. This erasure is accomplished by exposing the EPROM to ultraviolet light.

EXCLUSIVE OR If $F = A$ XOR B, then F will only be true if A or B but not A and B are true. In other words, F will be a one, if and only if one of the variables is true.

Flag A memory location used to indicate a particular condition. For example, an interrupt routine might set a particular bit in memory when a character has been received and is available for processing. This bit is called a flag.

Flip-Flop A special register that can toggle between a one and a zero state.

FPGA Acronym for field programmable gate array. Specifically, this term refers to Xylinx's RAM-based PLDs. The term is often used to refer to any large PLD.

FPLA Acronym for Field Programmable Logic Array. Signetic's term for the standard two-plane PLD (i.e., programmable OR array, programmable AND array).

Fuse Fuse, as it applies to programmable logic, refers to a section of metallization on an IC that can be opened by passing a large current through it. This allows the IC to store data by having the fuses selectively blown.

Gate The term *gate* is used to refer to some atomic combinatorial logic element. Typically, examples include OR gates, AND gates, and XOR gates.

Gate Array A gate array, as the name implies, is a collection of logically isolated gates on an IC. The gates are connected by adding a layer of metallization to the wafer, microscopically wiring the gates together. Gate array design is typically net-list driven as opposed to the logic driven PLDs.

High-Level Language A language that allows a greater degree of abstraction than an assembler. HLLs are generally machine independent and thus are more portable. HLLs are not typically used for programmable circuit design, since they require a relatively sophisticated set of resources to execute.

Interrupt An interrupt is a hardware event that causes a special routine called the *interrupt routine* to be invoked. Interrupts are typically caused by elapsed timers, characters being received, etc. The interrupt causes normal program processing to halt while the interrupt is executing.

Interrupt Routine An interrupt routine is a special software routine. What makes it special is that it is called whenever a certain hardware event (the interrupt) occurs. Interrupt routines are similar to subroutines.

Inverter A logic element that implements the NOT function. If the input of an inverter is low, then the output will be high. If the input is high, then the output will be low.

JEDEC Acronym for the Joint Electron Device Engineering Council, a body which produces a variety of standards for government and industry.

Jump A jump causes an instruction other than the next sequential instruction to be executed.

Latch A latch is used to hold (latch) the state of the data on its input. The most common type is the D-latch.

MAX Family A family of sophisticated PLDs produced by the Altera Corporation.

Maxterm When every variable in an expression (in either its complemented or uncomplemented form) is summed to form the function, the variables are called the maxterms of the function.

Microcode The binary bit patterns used by a microsequencer or a state machine.

Microcontroller A single-chip computer. A microcontroller contains the central processing unit, memory and peripherals necessary to realize a wide range of functions.

Microinstruction The microinstruction is one instruction in the microcode. The format of the microinstruction is defined by the microword.

Microsequencer A specialized state machine that executes microcode sequentially. Microsequencers typically have stacks, counters, and some limited branching capability.

Microword The microword is the bit pattern for the microinstruction.

Minterm When each variable in an expression appears exactly once (in either its complemented or uncomplemented form) and these variables are ANDed together to form the function, the variables are called the minterms of the function.

Mnemonics Mnemonics are symbols used to represent a certain bit pattern. For example, if the instruction to load an accumulator is instruction number 17_H, a mnemonic such as LDA might be used to represent 17_H.

Multi-Tasking Multi-tasking is a technique for allowing many different tasks to be executed in a controlled environment. Multi-tasking is used to improve the utilization of a computer's resources.

NAND The complement of the AND operation.

Native Code Native code refers to machine language instructions which are generated by the "native" machine. For example, if a PC is used to compile a program that will run on the PC, the resulting code is native code. See *target code*.

NOR The complement of the OR operation.

Object Code Object code is the machine level instructions for a computer. Object code is produced by compiling *source code*. The term *object code* is often used to describe the bit patterns produced by PLD assemblers or compilers.

Opcode The machine-readable bit pattern for an instruction. For example, if the mnemonic to load an accumulator is LDA, and the instruction number is 17_H, then 17_H is the opcode for LDA.

OR The OR operation yields a 1 if any binary variable has a value of 1.

PAL Acronym for programmable array logic.

Peripheral A peripheral is a device that is part of the CPU or memory. Examples of peripherals include timers, counters, UARTS, and I/O ports.

Pipeline Pipelining is a technique of adding registers into the feedback path of state machines or microsequencers. The purpose is to allow different operations to be accomplished at different points in the pipeline, thus improving overall efficiency.

PLD Acronym for programmable logic device. This is a generic term for anything that includes PALs, PLAs or special purpose devices.

PLE Acronym for programmable logic element. A term sometimes used to describe a ROM that is being used with a PLD.

Polling The process of repeatedly checking the status of an input. Opposite of interrupt.

Product of Sums An expression in the form of $(A + B) * (B + C)$.

Program A sequence of instructions and data for performing a specific function.

Program Store As the name implies, the place where the program gets stored. Typically a ROM, PROM, EPROM, or EEPROM.

Programmable Gate Array (PGA) A very sophisticated PLD with routing characteristics similar to a gate array.

Programmer A device used to transfer a program or bit pattern from the development station to a PLD, PROM, or microcontroller.

Pseudo-op Something that resembles an opcode but is not one. Examples include DB (define byte) and EQU (equate).

Pulse Width Modulation (PWM) A technique for controlling the average amount of power to a device by controlling the duty cycle of a switch.

RAM Acronym for random access memory. See Appendix A.

Register A convenient place to stash a value. Also used to manipulate values.

RISC Acronym for reduced instruction set computer.

ROM Acronym for read only memory. See Appendix A.

Security Fuse A special fuse that prevents the contents of a PLD from being read.

Software Any of the programs, utilities, or applications that are used by a PLD, state machine, or microcontroller.

Source Code The human readable input from which the object code is generated.

Stack A special area in memory that is used to store variables, return address, and machine status. The stack is controlled by the stack register.

State Condition of a finite automata at any given time. For a two bit counter the states are 0, 1, 2, and 3.

State Machine Loosely, any circuit that makes use of states. Typically used to define a circuit that makes use of a latch whose outputs are fed back to the input via a combinatorial network.

Subroutine A routine that is called by the main routine. Upon completion, the subroutine returns control to the instruction immediately following the calling instruction.

Sum of Products An expression of the form $F = AB + CD$.

Target Code Target code refers to machine language instructions which are generated by the "native" machine (the host) for use on the target machine. For example, if a PC is used to compile a program that will run on a microcontroller, the resulting code is target code. See *native code*.

Target Machine The machine that will be executing code being generated on a development system.

Up-Load To transfer data or programs from a target machine to the development system.

Wired-OR A technique for creating an OR-connection without actually using a gate. For example, the output of several open collector gates may be wired to a single pull-up resistor. Such a node is called wired-OR.

Writable Control Store A special type of program store for a microsequencer. A writable control store can be directly written to and is normally implemented with RAM circuitry.

XOR (Exclusive OR) $F = A$ XOR B is true if and only if A or B is true, but not both.

Index

Accumulator, 182
Active high, 17
Active low, 17, 20
AND, 18, 20, 22, 27, 50-52, 56
ASIC, 130-131
Assembler, 5, 92, 93
Assembly language, 202
Automatic test equipment (ATE), 151

Bandwidth, 251
Boolean, 13, 16
Boundary scan, 153
Branch instruction, 206
Built-in test (BIT), 153

Canonical form, 24, 26
CISC, 164, 174-175
Combinatorial logic, 19
Compiler, 5, 9
CPU, 69, 169-174, 181
Cross assembler, 179

Debounce, 33
De Morgan's theorem, 12, 16, 22, 26, 34
Development station, 3
Devices
  8048/8748, 191, 231
  8051, 191
  AM2901, 168
  EP-310, 131-132
  EPM5128, 135
  GAL16V8, 143-144
  ispGAL 16Z8, 141
  MAX Family, 131-138
  MC6801, 182, 191-192
  MC6805, 231
  MC68HC11, 183, 192, 220, 230
  PAL16L8, 56, 65, 99, 100, 104, 130, 143-144
  PAL16R4, 104, 106, 110, 115, 143-144
  PAL16R8, 143
  PAL22V10, 115-123, 130, 132, 133, 135-136
  PLHS-105, 123
  PLS-153, 57, 58, 63, 65
  PSG-506/507, 123
  SAM448, 168

EEPROM, 6, 162
EPROM, 6
Exclusive OR *see* XOR

Flip-flop, 29-32, 112
Floating gate, 37-39
FPGA, 138, 142
FPLA, 11, 36
Fuse, 37, 50

Gate, 19, 21, 112

In-circuit emulation, 244-246
Interrupt, 218-219
Inverter, 19, 21, 50, 56

JTAG boundary scan, 153
Jump, 204

Karnaugh maps, 28, 29

Latch, 29-30
Logic analyzer, 235-237

Maxterm, 25, 34
Mealy machine, 67-69, 75, 86
Metastability, 154-156
Microcode, 44, 67, 74, 202
Microcontroller, 1, 6, 15, 89, 177-181
Microinstruction, 171, 202
Microsequencer, 164
Microword, 74, 76
Minimization, 23

Minterm, 25, 34, 48
Mnemonics, 170
Moore machine, 67-69, 75, 86
Monitor, 237-241
Multi-tasking, 220, 233

NAND, 19
NOR, 19, 56
NOT, 22, 23

Object code, 90, 170, 202, 233
Opcode, 5
Optimization, 23
OR, 18, 22, 27, 35, 50-52, 56

PAL, 7, 8, 11, 14, 15, 35, 47-52, 54, 56, 57, 63, 65, 66, 82, 91
Pipeline register, 86, 168
PLA, 14, 15, 35-36, 47, 49, 52, 57, 63, 65, 66, 82, 123
PLD, 1, 3, 13, 14, 20, 29, 36, 38, 40, 44-52, 54, 59, 63-67, 69, 75, 89, 91, 92, 98, 99, 111, 131, 162, 180
  Development cycle, 3, 4
  Logic design, 13
PLE, 39
Polling, 210
Product of sums (POS), 25, 26, 34, 57
Programmer, 6
Programming packages
  ABEL, 95
  AMAZE, 94
  CUPL, 95
  PALASM, 94
  PLAN II, 96
  PLDesigner, 95
  proLogic, 97
PROM, 39-47
Pseudo-op, 42
Pulse width modulation (PWM), 197

Quine-McCluskey, 29

RAM, 59, 60, 177, 183
RISC, 164, 174-175
ROM, 177
RTL logic, 54

Security fuse, 162
Sequential logic, 13, 19, 29
Signature analysis, 153
Simulation, 108-111
Simulators, 241-244
Software *see* Programming packages
Source code, 5, 90, 170, 233
Stack, 205
State, 29
State diagram, 72, 76, 77, 79, 100
State machine, 1, 13-15, 67, 68, 76, 86, 251
State table, 72-74
Subroutine, 215, 218
Sum of products, 25, 26, 34, 47, 57

Target machine, 179
Test vectors, 109

WIRE-OR, 20
Writable control store (WCS), 175-176

XOR, 18-20, 22, 35, 57